Effects of Welding on Health III

An up-dated (June 1979-December 1980) literature survey and evaluation of the data recorded since the publication of the first report, to understand and improve the occupational health of welding personnel.

Research performed by the Franklin Institute under contract with the American Welding Society and supported by industry contributions.

By
Samir Zakhari
and
John Strange

Franklin Institute Research Laboratory, Inc.
A Subsidiary of the Franklin Institute
The Benjamin Franklin Parkway
Philadelphia, PA 19103

Prepared for:
SAFETY AND HEALTH COMMITTEE
AMERICAN WELDING SOCIETY
P.O. Box 351040
Miami, Florida 33135

International Standard Book Number: 0-87171-226-1
Library of Congress Number: 79-52672

American Welding Society, 550 LeJeune Road, Miami, FL 33126

This report is published as a service and convenience to the welding industry and is the product of an independent contractor (Franklin Institute Research Laboratory) which is solely responsible for its contents. The materials in this report have not been independently reviewed or verified and are only offered as information. AWS assumes no responsibility for any claims that may arise from the use of this information. Users must make independent investigations to determine the applicability of this information for their purposes.

Printed in the United States of America

Contents

Personnel

Authors of the report by Franklin Research Center were:

Samir Zakhari and John Strange

AWS Research Committee

A.N. Ward, Chairman	Caterpillar Tractor Company
K.L. Brown, Vice Chairman	Lincoln Electric Company
M.E. Kennebeck, Jr., Secretary	American Welding Society
J.S. Gorski	Kemper Insurance Companies
E. Mastromatteo	INCO Limited

Acknowledgements

The American Welding Society gratefully acknowledges the time and effort expended by the members of the Research Committee and the financial support of the program by industry contributions.

Supporting Organizations

Air Products and Chemicals, Inc.
Airco Welding Products
Allis-Chalmers
Alloy Rods Division, The Chemetron Corporation
AWS Detroit Section
AWS New Orleans Section
Arcos Corporation
The Binkley Company
Caterpillar Tractor Company
Chicago Bridge and Iron Company
Grove Manufacturing Company, Division of Kidde, Inc.
General Electric Company
The Heil Company
Hobart Brothers Company
Huntington Alloys, Inc.
Lincoln Electric Company
Miller Electric Manufacturing Company
National-Standard Company
A.O. Smith Corporation
Teledyne-McKay, Inc.
Trinity Industries, Inc.
Truck Trailer Manufacturers Association
Walker Stainless Equipment Company
Weld Tooling Corporation

Many other organizations have made contributions to support the ongoing program from May 1979 to the present.

Preface

This literature review has been prepared for the Safety and Health Committee of the American Welding Society to provide an assessment of current knowledge of the effects of welding on health, as well as to aid in the formulation of a research program in this area, as part of an ongoing program sponsored by the Committee. Previous work has included studies of fumes and gases, radiation, and noise generated during various forms of arc welding. Conclusions based on this review and recommendations for further research are presented in Section 5 of the report. Section 1 summarizes the occupational exposures. Sections 2 and 3 contain information related to the effects of exposure to byproducts of welding operations on humans and laboratory animals. Section 4 covers *in vitro* studies.

Referenced materials are available from the Franklin Institute.

Introduction

The American Welding Society (AWS) has been concerned about the possible health hazards due to the exposure of welders to fumes, gases, radiation, and noise from various welding processes. Although much has already been learned about welding processes and technology, the effects of welding on human health have not been fully understood.

To help the studying and understanding of the welding environment and its effects on health, the AWS has undertaken a literature review entitled "Effects of Welding on Health" that was published in 1979, and has been updated to cover the period of January 1978 to May 1979.

Since then, interest in studying the effects of the welding environment on the various physiological systems has continued and many articles have appeared in the published literature. The present effort represents a continuation of assessing the effects of welding on health and presents an update to include information published from June 1979 to December 1980.

The reader is cautioned that the papers reviewed were examined only for consistency. No independent checks of the experiments or findings were performed. This report must be read in conjunction with "Effects of Welding on Health" by Villaume et al., and the update by Zakhari and Anderson (Vol II). References are listed at the end in alphabetical order based upon first author.

Technical Summary

The objective of this report is to evaluate and present the state-of-knowledge of the effects of welding on health, to point out gaps in this knowledge, and to provide recommendations to the American Welding Society for future studies. This report covers the period of June 1979 to December 1980, and must be read in conjunction with the previous literature review (Villaume et al. 1978) and the update which covers the period January 1978 to May 1979 (Zakhari and Anderson, 1981).

This report, as is the case of the previous two reports, is divided into 4 sections entitled: 1) The Exposure, 2) Effects of Welding on Human Health, 3) Toxicologic Investigations in Animals, and 4) In Vitro Studies.

The Exposure

In this chapter, studies performed on welding fumes, gases, radiation, and noise are summarized.

Fumes

During welding, particulate matter is generated from the base metal, welding rod, the flux, any coatings or contaminants of the metal surface, or any combination of these. The rate of fume generation and the composition and particle size distribution of the fumes vary according to the metal being welded and the welding method used. The components that are of special concern to welders include silicon compounds, chromium, nickel, fluoride, copper and manganese. Because of the potential health hazard of hexavalent chromium to man, several efforts have been made either to minimize the amounts of chromium in welding fumes, or to develop sensitive methods of continuously monitoring its concentrations in the ambient atmosphere of welders and in their biological fluids.

Potential health hazards from welding fumes are determined not only by the composition of the fumes and their concentration in air, but also by the duration of exposure as well as the particle size distribution. Fortunately, these hazards can be overcome by the use of appropriate protective devices and by engineering controls.

Gases

Contrary to the common belief that ozone is found in high concentrations in the vicinity of welding areas, studies conducted by Press (1978) revealed that very low concentrations of ozone are formed during plasma arc cutting. These low concentrations were ascribed to the reaction of ozone with nitrous oxide and the disintegration of ozone due to the presence of dust particles.

Radiation

Welding produces electromagnetic radiation that can be divided into visible, ultraviolet, and infrared. Radiation of wavelength shorter than 175 nm are rarely encountered during welding due to their absorption by atmospheric oxygen (Lunau, 1967). The use of thoriated tungsten welding electrodes (GTAW process) could result in the exposure to thorium and daughters contained in these electrodes. The magnitude of the radiation to which welders may be exposed was estimated to be between 20 millirem and one rem.

Noise

Dixon (1978 a) found that the noise of an electron beam welding machine was directly proportional to the beam voltage and current.

Effects of Welding on Human Health

In this chapter, the effects of welding on human physiological systems are discussed. Data about carcinogenicity of welding fumes and gases, and epidemiological studies are summarized.

Toxicity to Various Organs

Respiratory System

Keskinen et al. (1980) reported on two asthmatic patients in whom attacks were precipitated by welding fumes. Examination of the occupational history of these patients revealed previous employment at a cement factory and an iron foundry where cement dust, clay dust, and iron fumes were prevalent. Chronic exposure to gases and particulates in welding atmosphere could result in interstitial fibrosis of the lungs (Loriot et al. 1979). On the other hand, Gola et al. (1980) found that only 7 persons out of a group of 73 arc welders, exhibited symptoms of chronic non-specific bronchopulmonary disease. Chronic bronchitis was also reported in welders with up to 30 years of exposure to welding fumes (Candarella et al., 1979). Other chronic effects reported included inflammatory changes of the mucosa (Kup, 1979), and pneumoconiosis (Rabenda, 1980).

Lung function tests have been used to assess the effects of acute and chronic exposure to welding fumes. However, results of the lung function tests were inconsistent and contradictory. For instance, Oxhoj et al. (1979) found that welders showed a significantly higher closing lung volume and capacity than the controls; on the other hand, McMillan and Heath (1979) found that welders showed obstructive lung diseases, whereas controls showed restrictive lung diseases. However, individuals with most abnormal patterns were smokers. The degree of contribution of smoking or welding fume to the condition is unknown. Although McMillan and Heath's study was well designed and executed, the statistical significance could vary by changing the size of the population. Ross (1978) examined 926 welders and controls for lung function tests and found that, on several occasions, welders had significantly better results than the controls.

Radiographic changes in lungs of welders have been studied by Attfield and Ross (1978) who found very few abnormalities until the worker had been employed for 15 years; the prevalence of pneumoconiosis steadily increased thereafter.

Ear and Hearing

Jarzebski (1979) reported on cases of middle ear burns due to sparks or foreign bodies.

Eyes and Vision

Alkali fumes produced during welding can cause considerable eye irritation. Exposure to ultraviolet radiation causes very little immediate discomfort on exposure but might, in severe cases, cause irreversible photochemical reaction in the pigment layer. Permanent damage could be caused by thermal effects which might be aggravated by the simultaneous photochemical reaction due to ultraviolet radiation.

Lovsund et al. (1979 b) found no evidence of any "alarming" effect on the visual system from short term exposure to magnetic fields generated during welding processes. Effects of welding on readaptation time (RAT, time that elapses from the moment of exposure to a bright light until a certain object reappears) was studied by Linde (1980) who found a marked increase in RAT with fumes from basic electrodes but not from rutile electrodes. Welders who wear contact lenses should use other means for adequate safety protection, and a qualified person should be available to remove the lenses in an emergency.

Skin

Most of skin injuries are caused by flying hot metal particles and exposure to ultraviolet radiation. Contact eczema was reported in unprotected electric welders (Weiler, 1979). An Epidemiological study of 200,000 cases of skin cancer failed to link this condition to welding.

Musculoskeletal System

An epidemiologic study was conducted by Nauwald in 1980 on 100 welders in the shipbuilding industry. Although the sample is too small to allow any significant conclusion, the author claimed that welders are at an increased risk of developing knee joint diseases.

Biochemical Effects

Heavy metals were found in various biological fluids and tissues of welders in higher concentrations than the control group. Grund (1980) found high manganese contents of hair samples obtained from shipyard welders; Bernacki et al. (1976) found high nickel concentrations in the urine of arc welders; exposure to welding fumes also resulted in a decrease in plasma proline, alkaline phosphatase, and lactate dehydrogenase; and an increase in plasma hydroxyproline (Mlynarczyk and Senczuk, 1980; Wysocki et al., 1980).

Mutagenicity and Carcinogenicity

Bloom (1979) studied the effects of ozone on human chromosome response in blood samples taken from 247 welding trainees. A slight rise in the percentage of cells with single chromatid and isochromatid gaps was demonstrated after 6 weeks of exposure, but declined by the twelfth week. However, no statistically significant increases were seen in any category of other chromosome-type abberations studied.

Epidemiologic Studies

Beaumont (1980) studied the mortality of welders and other trade workers in the Greater Seattle area. No definitive conclusions could be reached by Beaumont as to the cause and effect relationship of various diseases in welders; higher incidence of cancer of the respiratory system and pneumonia was found in welders. Since records of these welders showed that they performed various jobs (e.g. in shipyard, metal fabrication, field construction, etc.) in the same year, Beaumont's findings cannot be ascribed to a given occupational exposure.

McMillian (1979, 1980) conducted an epidemiologic study on the health of welders in the Royal Dockyards and found no evidence of an excess of chronic respiratory diseases.

Toxicologic Investigations in Animals

Rats were exposed to 1,178 mg/m^3 of fumes generated by SMAW process using flux coated electrodes for 45.8 minutes. The concentration of iron in the lung tissue was 2,185 ug/g of lung tissue immediately after exposure. This concentration was reduced to 1,085 ug/g 30 hours later, and 657 ug/g 30 days later. The overall elimination rates of iron was less than those of chromium and cobalt. However, the high concentrations of fumes used did not simulate the actual occupational exposure. For instance, in this experiment the average lung deposit in rat's lungs was 1.5 mg in 45.8 minutes (equivalent to 83 mg in human lungs); Kalliomaki et al. (1980) found an average deposit of 139 mg in welder's lungs after 9.9 years of welding.

Other studies performed in this period focused on the pattern of deposition and the elimination rate of fume particles from lungs (Lam et al., 1979; Alshamma et al., 1979 a,b).

In Vitro Studies

In 1980, Knudsen used the mammalian spot test to study the mutagenicity of fume particles. Knudsen found that intraperitoneal injection of pregnant female mice with 100 mg/kg of fume particles, collected from SMA welding of stainless steel, resulted in the development of grayish spots in the fur of offspring. The dose used seems to be very large, and the fact that the intraperitoneal route of administration does not represent the inhalation route by which welders are exposed to fumes. Therefore, the relevance of this study to the effect of welding on health is very questionable.

Effects of Welding on Health III

1. The Exposure

Welding is a materials joining process which produces coalescence of materials by heating them to suitable temperatures, with or without the application of pressure or by the application of pressure alone, and with or without the use of filler metal. During this process a welder is exposed to different factors which, in the absence of appropriate protective measures, might affect his or her health. These factors include the production of fumes, gases, radiation, heat, and noise. The potential hazards are well-known, and by exercising precautions and alertness, the possibility of endangering the welding operator is remote. Hazards associated with welding can be overcome by the appropriate use of protective devices and engineering controls (Eaton, 1977). Fumes generated during welding pose least hazard to welders compared to burning, cutting or mechanical injury (Tierney, 1977), although it is of major concern to industrial hygienists. This update of the effects of welding on health which covers the period of June 1979 to December 1980, will be organized as previously submitted reports (Villaume et al., 1978; Zakhari and Anderson, 1979).

1.1 Fumes

There are four major factors that influence the level of fumes and level of risk to the welder's health. These factors are:

A) The welding process, current settings, and the metal being welded. This will affect the quantity of the fumes, their composition, and particle size distribution: the AWS laboratory test method (AWS F1.2-1979) is recommended for the collection of fumes and for determining the composition and the generation rates of fumes produced during welding.
B) The duration of exposure to fumes and environmental aspects (Boekholt 1977 a,b).
C) The use of personal protective equipment and safety engineering.
D) Personal work characteristics such as posture, and speed of work (Evans et al., 1979).

Particulate matter and gases are generated from the base metal, the welding rod, the flux, any coatings or contaminants of the metal surface, or any combination of the above. Over 80 different welding processes in commercial use have been identified by the American Welding Society (1976). However, the most widely used processes are the following:

1) Arc welding, including
   - Shielded metal arc welding
   - Flux-cored arc welding
   - Submerged arc welding
   - Gas metal arc welding
   - Gas tungsten arc welding
   - Plasma arc welding
   - Arc stud welding
   - Percussion welding
2) Electroslag and electrogas welding
3) Resistance welding
4) Flash and friction welding
5) Electron beam welding
6) Oxyfuel gas welding
7) Thermal spraying

For details of these methods, the reader is referred to technical publications of the American Welding Society and others.

In the case of arc welding, the rate of fumes generation is dependent upon the welding electrode, the welding current, the arc voltage, the electrode polarity and diameter, and welding practices (Akbarkhanzadeh, 1979). Fumes are generated at a high rate in the flux-cored arc process. The oxyfuel gas process generates a significant amount of fumes only when welding galvanized steel or aluminum. New interest has been expressed in the submerged arc welding process, in use since the 1930's, (British Patent, 1976; Prosperi, 1979; Wilkinson, 1979; Pinfold et al., 1979; Weymueller, 1979a,b; Post, 1980), and other methods have been developed for electric spot welding and for totally confined explosive welding (Bement, 1978).

The rate of fume generation, the nominal hygienic air requirement, and the threshold limit value index for the electrodes are the three major ciriteria for the classification of electrodes in Sweden. Each of these values has been determined for a variety of electrodes by Miller and Jones (1979). Gobbato et al. (1979) found that all electrode coating components are present in the welding fumes to a varying degree; however, three types of characteristics were shown for covered electrodes with carbon steel core wires: a) elements that have similar concentrations in the fumes and coat, which include iron, magnesium, and calcium, b) elements that are more concentrated in the fumes, which include manganese, copper, zinc and lead, and c) elements that show lower concentrations in the fume include chromium and nickel. Fewer fumes were found to be generated from flux-cored electrodes or electrodes in which the core or filling contained larger than normal amounts of certain basic oxides and limited amounts of other fluxing ingredients such as acids, amphoteric oxides, or fluorides (Frew, 1979). Kobayashi et al. (1979) claimed that seven new electrodes have been developed that have low emission rates. The chemical composition of fumes and the mechanical properties were given.

The composition and particle size distribution of the fumes vary from one metal to another, and as previously mentioned, depend on the welding method used. Some components of the fume may pose more potential health hazards than others do. The relative hazard depends on the form of the component present and its concentration in the air vs. the concentration of other components in the welding environment. Components that are of special concern to welders include silicon compounds, chromium, nickel, fluorides, copper, and manganese.

Silicates are present in the coating of shielded metal arc electrodes and the core of flux-cored arc electrodes. Only the crystalline forms of silicon dioxide are responsible for the induction of silicosis. However, crystalline phases of silica have not been found in welding fumes (Buckup, 1973; Patter et al., 1978).

Chromium, in the form of hexavalent, trivalent, and divalent states, is present in the fumes produced by welding stainless and high alloy steels.

Mutagenicity and carcinogenicity of chromium compounds have been associated only with the hexavalent form (NIOSH 1975; Maxild, 1978). Stern (1977) found that the concentration of hexavalent chromium in welding fumes varies not only with the nature of the metal being welded but also with the method of welding itself. Thus the hexavalent chromium concentration was higher in fumes produced by welding stainless steel than those from welding mild steel; gas metal arc welding yielded higher chromium content than shielded metal arc welding. These findings were in contrast to those of Virtamo and Tuomola (1974), who found that the gas metal arc process produced fumes with lesser hexavalent chromium content than those produced by the shielded metal arc process.

Because of the potential health hazard of hexavalent chromium to man (Langard and Norseth, 1975), several efforts have been made either to minimize the amount of chromium in welding fumes, or to develop sensitive methods of continously monitoring the chromium concentrations in the ambient atmosphere of welders and in their biological fluids. Furthermore, the position of the American Welding Society in relation to occupational exposure of welders to chromium is explained in a letter to NIOSH (DeLong, 1979). In that letter, AWS challenged the sensitivity of the method of chromium determination advocated by NIOSH. This opinion was shared by Thomsen and Stern (1979), who recommended the use of a carbonate leaching technique for the determination of both soluble and insoluble chromium.

A U.S. Patent by Kimura (1978) describes a coated welding electrode that contains at least 0.5% of chromium and is composed of a metal core and a coating flux composition where the contents of sodium and potassium components is reduced below 1% based on the total weight of the coating flux and lithium compounds are used in the coating. When welding is carried out using these electrodes, soluble chromium in the generated fumes is claimed to be substantially reduced.

Efforts to develop selective and sensitive methods for the determination of hexavalent chromium are triggered by the fact that the current OSHA Permissible Exposure Limit (PEL) is expected to be substantially lowered to 1 ug/m3 (NIOSH, 1975). A method for the separation and determination of trivalent and hexavalent chromium in welding fumes was described by Naranjit et al. (1979). In that method anion- and cation-exchange resins were used to separate both forms of chromium, and quantification was then carried out by the use of atomic absorption spectrometry. Many potential interferences were found to be present in the aqueous extract of welding fumes. The use of a very fuel-lean flame to overcome this problem resulted in a five-fold loss in sensitivity.

An analytic procedure for the determination of chromium in welding fumes advocated by Bohgard et al. (1979) involves four steps: a) total chromium is measured by the use of proton-induced X-ray emission; b) the oxidation state of the particle surface is measured by electron spectroscopy for chemical analysis; c) water-soluble hexavalent chromium measured by electron microscopy is used to provide information about the size and shape of the particles.

Infrared spectrometry was used by Kimura et al. (1979) to detect hexavalent chromium in solid fume samples. The authors used this method to compare the composition of fumes obtained by two welding methods. Fumes produced during welding of stainless steel using the shielded metal arc method contained 3% - 7% chromium, 60% - 90% of which is water soluble (hexavalent chromium). Fumes generated using gas metal arc welding contained 15% chromium, of which water soluble hexavalent chromium constitutes a very little fraction (unspecified).

Nickel is usually present in fumes produced by welding stainless steel and nickel alloys.

Fluorides in welding fume are mainly generated from the covering on shielded metal arc electrodes and the core of flux-cored arc electrodes. Fluoride compounds constitute 5%-30% of the total fume produced from basic covered electrodes and are present mainly as sodium, potassium, and calcium fluorides.

Copper is present in fumes produced from welding copper alloys and to a lesser extent from copper-coated gas-metal arc electrodes. Studies by Shutt (1979) showed that electrodes containing 0.18% copper did not result in excessive copper in the fumes.

Manganese, present in the fumes mainly as manganese oxides, is present in the coating of some shielded metal arc electrodes, and in the core of flux-cored arc electrodes (Pattee et al., 1978). Steels containing a high manganese content are another source of oxides of manganese in the welding fumes (Moreton, 1977).

Other components of welding fumes such as zinc (National Safety Council Data Sheet, 1978) cadmium (Hughes, 1980; King, 1980), cadmium oxide (Kaplan et al., 1977), and beryllium (Heiple and Dixon, 1979) were also discussed.

Potential health hazards from welding fumes are determined not only by the composition of the fumes but also by the particle size distribution. The latter is an important factor because it determines the locus in which various particles are deposited in the respiratory tract. The bulk of particles that range in size from 0.5 to 0.7 um could reach the alveoli; larger particles are mainly trapped in the upper respiratory passages. Clapp and Owen (1977) found that an appreciable amount of welding fumes is in the 1 to 7 um diameter range, whereas Stefanescu et al. (1968) found that 50% of the particles are of a size less than 0.1 um.

Others found the welding fume particles to be essentially less than 1.0 um in diameter (Akselsson et al., 1976; Hedenstedt et al., 1977). Clapp and Owen (1977) ascribed the larger size of particles to agglomeration and size growth of the particles by time. Particles deposited in the respiratory system could induce pulmonary irritation or systemic toxicity (lead, chromium, nickel, fluorides, zinc, titanium, vanadium, and manganese), pulmonary fibrosis (crystalline silica), or simply pneumoconiosis (iron and aluminum).

1.1.1 Rate of Fume Generation

Measurement of the fume emission rates of arc welding is important for the investigation of various factors including ventilation requirements, welder exposure, and fume composition. Various methods used to measure the rate of fume generation are gravimetric and depend on the weight of a filter used to trap certain amounts of the fumes. Details of various methods adopted by British, Swedish, Japanese, and American scientists are mentioned in articles by Oakley (1979) and Gonzales and Gomez (1979). Other methods of sampling were discussed by various authors (Wilson et al., 1981; Krieg, 1979; and Magnusson, 1977). Furthermore, measurements of lung contamination with welding fumes were accomplished by Kalliomaki et al. (1980 a,b).

Efforts to reduce the amount of fume in the welding environment were concentrated in three areas: a) the use of electrodes that produce less fumes (Kobayashi et al. 1979; Miller and Jones, 1979; Riedinger and Beaufils, 1978; Gonzales and Amato, 1979), b) recirculation of industrial exhaust air without loss of heat by using an electrostatic air cleaning system (Schubert, 1979) or other system (Hagopian, 1980; Anonymous, 1980; Sterling, 1979), and c) engineering measures designed to reduce the amount of fume in the breathing zone (Anonymous, 1979 a; Birchfield, 1980 a, b; NIOSH, 1979; Head and Silk, 1979; Honavar, 1979; Van Wagenen, 1979; Bartley et al., 1980; Astrop, 1980).

1.2 Gases

Gases generated during welding have their origin in

- atmospheric gases, e.g., ozone and nitrogen oxides
- shielding gases, e.g., argon and CO_2
- decomposition of solvents and coatings on surfaces

Studies conducted by Press (1978) to determine the concentrations of gases in plasma arc cutting revealed that very low concentrations of ozone are formed during welding. These low concentrations were ascribed to the reaction of ozone with nitrous oxide and the disintegration of ozone due to the presence of dust particles. It was claimed that by proper selection of working conditions (such as cutting speed, current, and plasma gas mixture), it is possible to keep the concentrations of noxious substances within low limits.

The formation of nitric oxides in gas welding and cutting processes is influenced by several factors. It was shown that the torch size and the distance of the torch from the workpiece are the main factors that affect the formation of nitric oxides (Commission VIII Health and Safety of IIW, 1980). Similar findings were previously reported by Ulfvarson et al. (1978).

1.3 Radiation

Welding is known to produce electromagnetic radiation which can be divided into visible, ultraviolet, and infrared. The degree of radiation exposure differs according to the welding process used and the metal being welded. The known human biological effects of radiation are summarized in Table 1-1 (NIOSH 1978). The adverse effects categorized by wavelength are shown in Table 1-2 (Hinrichs, 1978). Hinrichs (1978) found that gas metal arc welding produced the most radiation and plasma arc welding the least radiation. The intensity of radiation was found to be parallel to that of the welding current. Bartley et al. (1979) found

Table 1-1
Effect of electromagnetic radiation exposure

Microwave and radio-frequency radiation	Optical radiation	Ionizing radiation
Genetic effects	Skin cancer	Cancer
Teratogenic	Squamous cell epithelioma	Leukemia
Behavioral changes	Basal cell epithelioma	Sterility
Cataract	Herpes Simplex lesions	Cataract
Headache	Cataract	Radiodermatitis
Nervous system disorders	Retinal damage	Ulcerations
Loss of memory	Genetic alteration	Erythema
Dullness	Erythema	Genetic alteration
Sleepiness	Kerato conjunctivitis	Blood cell changes
Sinus arrhythmias	Heart stress	Malaise
	Cutaneous burns	Fever
	Asthenopia	Nausea
		Weakness
		Shock
		Diarrhea
		Life span shortening
		Anorexia

Table 1-2
Adverse biological effects due to radiation

Radiation	Wavelength (um)	Adverse biological effects
UV-C	100-280	Arc flash and slight erythema
UV-B	280-315	Arc flash and slight erythema
UV-A	315-400	Arc flash, erythema, cataracts, and skin cancer
Visible	400-700	Photochemical or thermal retinal injury, and thermal skin burns
IR-A	700-1,400	Retinal burns, thermal skin burns, and cataracts
IR-B	1,400-3,000	Thermal skin burns

that the presence of magnesium in aluminum alloys led to the formation of considerably higher radiation levels than did non-magnesium alloys. This was ascribed to the ease of vaporization of magnesium into the arc.

Thoriated tungsten welding electrodes containing 1%-2% radioactive thoria were introduced in 1952 (Breslin and Harris, 1952; Gibson and Seitz, 1952) for a variety of arc welding situations. The advantages of using thoriated tungsten rather than tungsten are ease of starting, less weld metal contamination, and greater arc stability. Welders are potentially exposed to the thorium and daughters contained in these electrodes. The magnitude of the radiation to which welders may be exposed was estimated to be between 20 millirem and 1 rem for a bone dose (2.4 - 88 millirem whole-body dose) (McDowell-Boyer, 1980 a,b). Welders who may receive such doses were assumed to use thoriated electrodes for an average of 100 hours/year in a shop (4 hours/day) and 200 hour/year welding at home. Since this is a typical situation, a more realistic estimate based on welding in shop only, was calculated. A bone dose was estimated to range between 0.9 and 160 millirem.

Radiations of wavelength shorter than 175 nm are rarely encountered during welding due to their absorption by atmospheric oxygen (Lunau, 1967). The intensity of ultraviolet radiations are also attenuated by the presence of welding fumes (Emmett and Horstman, 1976).

Research performed by NIOSH (1978), Campbell (1979), and Moss (1978, 1980) to determine the degree of protection against radiation offered by transparent curtains, showed that a gray curtain would appear to be the best choice. In almost every curtain tested, wavelengths of less than 340 um were greatly supressed. This means that most transparent curtains would give adequate protection in the UV region. Yellow curtains appeared to give better visibility of the workplace.

1.4 Noise

Welders are invariably exposed to noise that is produced not only by welding processes but also that produced by other operations in the workplace. The permissible 8-hour noise exposure level established by OSHA is 90 dB. Dixon (1978 a) studied the noise of an electron beam welding machine and found that noise is directly proportional to the beam voltage and current.

The overall measures that ensure the safety of welders in the workplace have been discussed in detail (OSHA, 1976; National Safety Council Data Sheets 1965, 1978 a,b, 1980). The use of welding curtains (Shumay, 1978) to reduce radiations and the use of safety glasses (Sherr, 1980; Horstman and Ingram, 1979) were discussed elsewhere. The OSHA standards for safety and welding have resulted in various authors discussing their merit (Betz, 1978, 1979; Speight, 1978; Weymueller, 1979, 1980 a,b; Westgate, 1979).

2. Effects of Welding on Human Health

Recently, several reviews have been published on the effects of welding on human health (Ross, 1978; Sloan, 1978; Villaume, et al., 1978; Zakhari and Anderson, 1979). This update of the effects of welding on health which covers literature published during the period of June 1979 to December 1980 will follow the same pattern of the earlier report (Zakhari and Anderson, 1979).

2.1 Toxicity to Various Organs

The effects of Welding fumes on various physiological system will be discussed.

2.1.1 Effects on The Respiratory System

During welding, a variable amount of gases and fumes are present in the welding area. In the absence of adequate ventilation and if protective equipment is not used, these gases and fumes may find their way to welders' lungs, where they can precipitate acute or chronic effects depending on the concentration and the duration of exposure. The gases normally present in welding environments include nitrogen oxides, ozone, carbon monoxide and carbon dioxide. Fumes, or suspended particulates, that reach the lungs might produce pneumoconiosis or pulmonary irritation and systemic toxicity.

Acute respiratory insufficiency might be precipitated by fumes and gases (metal fume fever will be discussed concomitantly with respiratory effects because lung damage is presumed to play an important role in the etiology of fever). Chronic obstructive airway diseases could be further enhanced or complicated by smoking tobacco or non-tobacco products (Antti-Poika et al., 1977).

In order to detect early any effect of welding fumes on welders' lungs and to screen potential welders (preemployment examination), several methods are used to assess respiratory function. These methods include a preliminary medical examination, respiratory questionaire, clinical examination (the frequency of occurrence of certain lung diseases), lung function tests, and radiography. The discussions that follow will be arranged under the following topics: acute respiratory diseases, chronic respiratory insufficiency, and methods of detecting respiratory effects.

2.1.1.1 Acute Respiratory Diseases

For a given inhalant to produce lung injury, it must be 1) in the form of very small particulates that are capable of reaching the lower respiratory tract, 2) deposited or absorbed on the bronchial or alveolar surfaces, and 3) retained long enough to induce damage (Jones and Wiell, 1978). This is true with all agents, whether they exist as gases or particulates. Allergic reactions also might occur and result in the precipitation of asthma. Keskinen et all. (1980) reported on two such asthmatic patients in whom attacks were induced by welding fumes. One patient had an earlier contact dermatitis due to chromium. Examination of the occupational history of that patient revealed 6 years of mild steel welding, preceded by 4 years in the concrete products industry, where he was exposed to cement dust containing chromium. He had also spent 6 years in an iron foundry where sand and clay dust and iron fumes are prevalent. This patient experienced a delayed type of asthmatic reaction after welding stainless steel. The other patient, 50 years old, spent 25 years in mild steel and stainless steel welding and showed an immediate type reaction. Asthmatic attack in both patients was prevented by pretreatment with disodium chromoglycate and beclomethasone. Provocation with fumes from mild steel welding failed to induce any asthmatic response. The authors ascribed the effect to the chromium and nickel content of fumes from stainless steel welding.

2.1.1.2 Chronic Respiratory Diseases

Fumes and gases generated from the welding processes can gain access to the respiratory system of welders especially in the absence of proper protective measures. These fumes and gases contain elements or compounds that could be harmful to health. An understanding of what fumes and gases are present in each welding process and consumable, and under what welding conditions or plant environment, is important. The following paragraph lists most of the components which should be considered. Most of the components are present only under special conditions and only in a few limited consumables and processes.

Gases are classified as primary pulmonary irritants such as ozone, nitrogen oxides, phosgene, and phosphine or nonirritants such as carbon monoxide and carbon dioxide. Particulates are subdivided into those that produce pneumoconioses such as copper, beryllium, tin, iron, and aluminum and those that produce pulmonary irritation such as cadmium, chromium, lead, fluorides, manganese, mercury, magnesium, nickel, molybdenum, titanium, vanadium, and zinc (Weymueller, 1979). Using proton-induced X-ray emission, Akselsson et al. (1976) analyzed the particle size distribution and human respiratory deposition of trace metals. They found that for respiratory deposition, there was no significant difference between various elements. After deposition, the defense mechanisms of the respiratory membranes were activated to rid the body of these foreign elements. These defense processes are depicted in Figure 2-1 and are detailed in a review by Green et al. (1977).

Continuous exposure to these gases and particulates could result in the overwhelming of the clearance mechanisms and hence chronic respiratory diseases. Loriot et al. (1979) reported a case of an arc welder with diffuse interstitial fibrosis of the lungs. The case involved a 43-year-old welder who had spent about 10 years in arc welding. There appeared to be negligence on the part of the welder in using protective measures. Insufficiency of the right ventricle of the myocardium might ensue in such cases. The examination of a group of 73 arc welders by Gola et al. (1980) revealed that only 7 persons, all under the age of 50, showed symptoms of chronic non-specific bronchopulmonary disease. However, none of these patients showed any significant change in serum alpha-I-antitrypsin. These observations are in support of findings by Candarella et al. (1979), who examined 95 welders with up to 30 years of exposure to welding fumes. Results revealed a high incidence of chronic bronchitis and of impairment of lung function, mostly the obstructive type. Cytological examination of sputum showed squamous

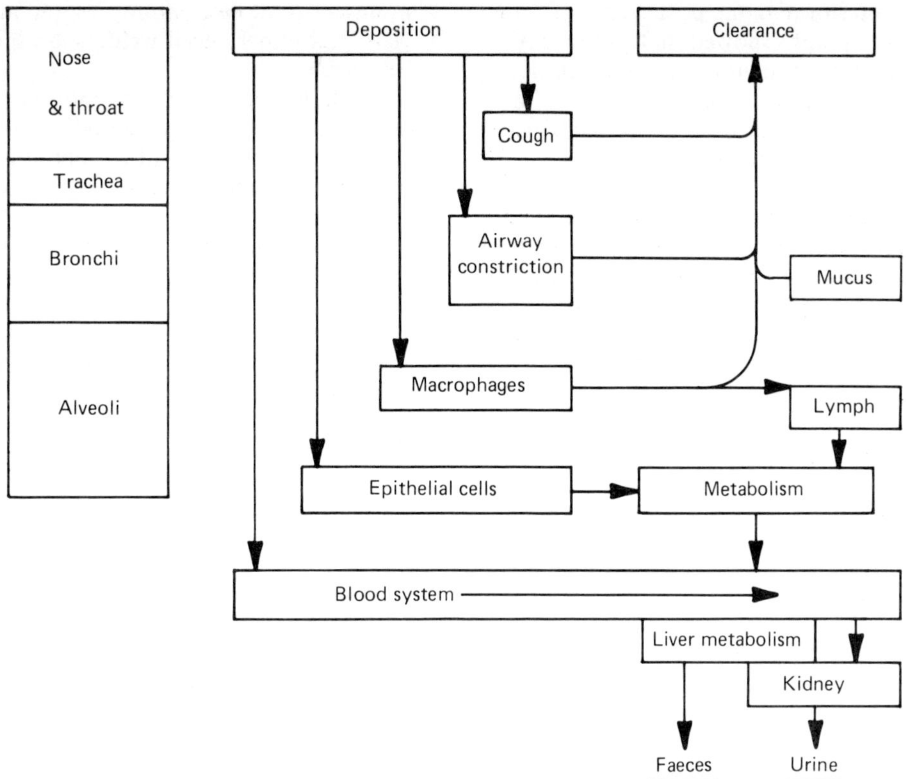

Fig. 2-1 — Pathways of elimination of fume dust

metaplastic cells possibly due to the presence of alleged carcinogens in the welding fumes.

Funahashi et al. (1979) used energy dispersive X-ray analysis (EDXA) to analyze lung biopsies of patients with welders' pneumoconiosis. Pulmonary siderosis commonly seen in welders is usually non-fibrotic with minor clinical symptoms, and histologic examination reveals iron particle-filled macrophages. In case with significant clinical symptoms, there is generally considerable interstitial fibrosis; it has usually been assumed that inhaled crystalline silicon dioxide caused the fibrosis. However, this study found that the technique of EDXA failed to show a higher than normal content of silicon dioxide compounds in the biopsied lungs examined in 8 cases of pulmonary siderosis. The responsible agent was not identified. Kup (1979) studied the occupational effect of welding on the upper respiratory tract mucosa of welders and found that 70% of workers examined had inflammatory changes of the mucosa.

Particulate matter is usually present in welders' lungs, especially in the absence of proper ventilation, and may lead to pneumoconiosis. Rabenda (1980) found no significant difference between the prevalence of pneumoconiosis among welders operating automatic or semiautomatic systems. However, the epidemiological study performed by Rabenda did not specify the exact conditions of exposure and whether these welders were previously employed in other "dust"-producing jobs. The exact amount and distribution of fume contaminants in the lungs of a 35-year-old male arc welder was investigated by Kalliomaki et al. (1979). The arc welder had worked for 11 years in a shipyard welding mild steel plates using mainly basic coated electrodes. He had never smoked and had been in good health without respiratory symptoms until he died suddenly of myocardial infection. His lungs were removed and investigated by morphological methods, including gross pathology radiography, histological and ultrastructural examination, chemical analysis of lung tissues, and sensitive magnetic measurements. The technique of the measurement of the magnetic lung contaminants was later used for the *in vivo* measurements, and details were given by Kalliomaki et al. (1980 a,b). The amount of welding fume contaminants found was 110 mg, of which 10% was iron. This amount is much less than the average value of 700 mg reported by Kalliomaki et al. (1978) for welders in the same shipyard. No explanation was given. Contaminants tended to collect in two concentration centers of the lungs. One anterior and one posterior. Transmission electron micrographs showed that enlarged lysosomes of macrophags contained electron-dense granules.

In contrast to the magnetic field measurements used by Kalliomaki et al., Kujawska and Marek (1979) used simple radiological examination to detect pneumoconiosis in welders' lungs. Radiological examination of 35 welders with pneumoconiosis showed that changes in welders' lungs may be reversible. Cessation of exposure to welding fumes resulted in complete remission of radiological changes in 9 cases, clear reduction in 16 cases, and no change in the other 10 persons.

2.1.1.3 Preliminary Medical Examination and Respiratory Questionnaire

Although a periodic comprehensive physical examination of welders would have far-reaching effects on occupational health, it is argued that greater benefits for welders and their employers would be achieved if routine clinical examinations were replaced by health screening (McMillan, 1979). This hypothesis is based on the fact that singling out certain employees for more careful medical examination might be interpreted by these employees as an indication that they are being exposed to more health hazards than others. Moreover, acute effects such as metal fume fever and arc eye, that are due to negligence in the use of protective measures and hence overexposure to fumes, gases, and radiation, will be neither detected nor prevented by a comprehensive medical examination. Even with the strictest adherence to threshold limit values, there is a small percentage of welders who are more sensitive to compounds in fumes than others and who might be adversely affected by low concentrations. Welders who have pre- or co-existing chronic obstructive airways, especially tobacco smokers, might be overlooked in a routine medical examination. McMillan (1979 a) maintained that welders required health surveillance designed specifically to detect obstructive airways disease. The suggested scheme is shown in Figure 2-2. The scheme is based on the detection of chronic lower respiratory tract diseases before and each year following employment. This is carried out by 1) a pre-employment or an initial health interview, 2) laboratory determination of lung function test, 3) ongoing monitoring, 4) review of sickness absence, and interviews on return from absence attributed to lower respiratory tract.

Laboratory examination includes dynamic lung volumes (e.g. FEV and FVC) and eye-to-work distance. The latter is important because welders with defective vision may have to bring their eyes and hence the breathing zone closer to the plume than they would if their vision were normal or properly corrected.

McMillan (1979) suggested that the frequency of administering the questionnaire be a period of three months for all new employees and at the end of first year. For experienced welders, an interval of 5 years was suggested.

Keskinen et al. (1980) reported on occupational asthma due to fumes generated from welding stainless steel by the manual metal arc method. One patient had had earlier contact dermatitis due to chromium, and another patient was atopic, since he showed an immediate type reaction. Chromium and/or nickel in the welding fumes was described as the etiological factor for the precipitation of asthma. The role of various factors in the genesis of asthma and different lung diseases was discussed by Jones and Weill (1978) and Carta and Sanna-Randaccio (1977).

2.1.1.4 Lung Function Tests

Lung function tests have been used to assess the effects of acute (Anthony et al., 1978; McMillan and Heath, 1979) and chronic (Ross, 1978; Akbarkhanzadeh, 1980) exposure to welding fumes. While these tests are commonly used in clinical medicine, significant deviation from the so-called "normal" values might occur in advanced cases, especially those with restrictive or obstructive lung diseases. Investigators reported conflicting results in lung function. Unfortunately, no conclusion can be reached on the basis of these studies. Variations in study design and parameters measured prevent direct comparison or accumulation of results; for instance, co-existing or previous exposure to asbestos was considered only by Ross (1978) and McMillan and Heath (1979). Different lung function parameters were studied by various investigators, and it is not clear whether changes in lung function were due primarily to welding fumes or to smoking.

Welding fumes are comprised of various gases and particles that range in diameter from 0.01 to 1 um (Villaume et al., 1978). Particles of that size are usually deposited in the peripheral non-ciliated broncheoli and in alveoli (Task Group on Lung Dynamics, 1966; Lippmann, 1972). If welding fumes are hazardous to the respiratory system, it is expected that pathological changes will presumably be located mainly in the peripheral airways and in the alveoli. Since resistance to air flow is very small in the peripheral airways, constituting as low as 10% of the total airway resistance, obstruction located in these peripheral airways could not be detected by conventional spirometry. It is, therefore, not surprising to have conflicting results of spirometric studies. Changes in small airways could be measured by the simple single-breath test that was introduced a few years ago (McCarthy et al., 1972; Buist et al., 1973; and Oxhoj et al., 1977). Oxhoj et al. (1979) used this method to study the lung function of 119 electric arc welders who spent 5-38 years at a shipyard where new ships were being built. Welding was mainly performed indoors using HT steel and basic electrodes (similar to AWS 7018). Half of the

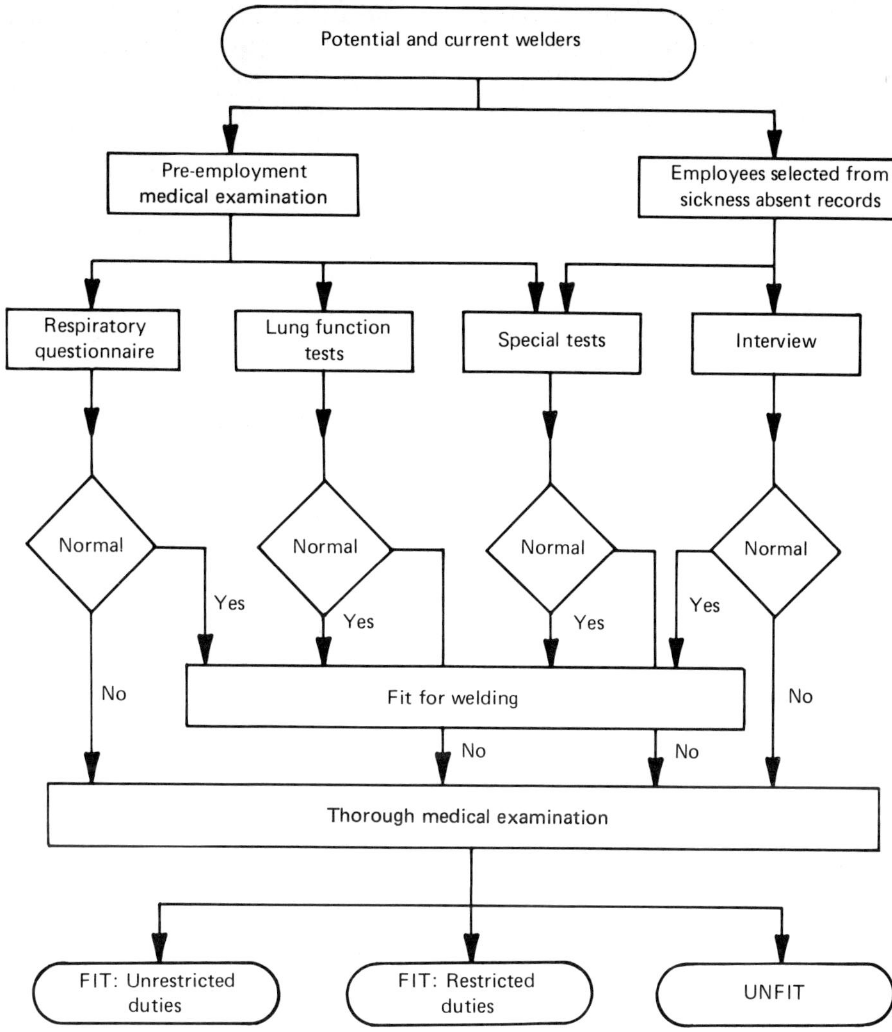

Fig. 2-2 — Scheme of health surveillance of welders. Modified from McMillan (1979)

welders worked in a mean total dust concentration exceeding 10 mg/m^3. Analysis of dust composition by energy-dispersive Roentgen-Fluorescence showed a higher concentration of copper, lead, gallium selenium, rubidium, strontium and potassium in particles of the order of 0-1 m in size. In comparison with the mean of 90 controls, welders showed a significantly higher closing lung volume and capacity. The total lung capacity was significantly lower in the welders who were smokers or ex-smokers. These findings were attributed to the deposition of welding fume particles in peripheral small airways or alveoli.

In contrast to the Oxhoj study, McMillan and Heath (1979) studied the acute changes in respiratory function of 25 welders with 6-25 years of experience during which the exposure was monitored. A group of 25 electrical fitters matched to welders by age, sex, and smoking habit was used as the control group. Each participant from the welders and the control group was clinically examined, had a chest radiograph, and performed a number of pulmonary function tests at the beginning and end of a shift. Each welder worked for 1 day, giving a total of 25 separate test days over a 5-week period. Each welder worked on welding the seams of dock blocks using general purpose rutile electrodes (4 SWG Vordex) in a box-like metal structure with only natural ventilation; no respiratory protection was worn. McMillian and Heath's study was designed such that it eliminated several uncontrollable variables encountered in a random sample. However, acute affects could be overshadowed by superimposition on chronic effects of welding.

No significant differences in the frequency of occurrence of respiratory signs and symptoms on clinical examination were found between welders and controls. Nine welders (six smokers) and eight controls (seven smokers) reported one or more respiratory symptoms as follows:

Symptom	No. of welders	No. of controls
Chronic bronchitis	5	2
Rales or ronchi	5	3
Absent from work for more than a week due to respiratory disease	8	11
With radiographic abnormalities	9	4

No siderosis could be detected in welders. However, eight welders and two controls showed plural abnormalities that were ascribed to previous exposure to asbestos.

Lung function tests revealed no significant difference between the overall prevalence of abnormal patterns (13 welders and 14 controls). However, qualitative differences between welders and controls were observed; the degree of obstruction was greater in the welders, and the degree of restriction was greater in the controls. Most of those with abnormal patterns were smokers. Although acute changes in lung function reflected no significant differences between the group means of welders and controls or those of smokers and nonsmokers, a positive correlation between changes in residual volume and total fume, respirable fume, iron oxide fume, and NO_X was found. This correlation could be ascribed to obstructive changes in the small airways.

Although McMillan and Heath's study is well designed and executed, the statistical significance could vary by changing the size of the population studied. Extrapolation from this study to welders in general could be misleading. At least theoretically, welding fumes and gases contain a mixture of bronchial and alveolar irritants which could give rise to obstructive lung diseases, especially in the smokers.

Spirometric examination of 1,232 of workers exposed to dust and welding smoke (Juniczak, 1979) showed a decrease in the ventilation efficiency of lungs of 39 workers (3.2%) with at least 16 years of exposure. Of these workers, 38 showed an obstructive pattern of ventilation abnormality. Concomitant exposure to asbestos and history of smoking were not given. These results are in sharp contrast to those of Ross (1978) who examined 926 welders and controls for Forced Vital Capacity (FVC), Forced Expiratory Volume for one second (FEV 1.0) and the Peak Expiratory Flow (PEF). In fact, he found that on several occasions, welders had significantly better results than the controls. The only group of welders that showed restrictive ventilatory capacity compared with the controls were those who smoked and were in the 40-49 age group. Overall, 8.4% of smoking welders and 6.6% of non-smoking welders had some restrictions to their ventilatory capacity.

To study the long-term effect of welding fumes and of cigarette smoking on the respiratory system, Akbarkhanzadeh (1980) compared the spirometric measurements of 209 welders with those of 109 nonwelder control. The two groups were matched for age, height, smoking habits, residence, and social class. Results showed that the prevalence of respiratory symptoms increased with age and was greater among the welders than among the controls, and was greater among smokers than among non-smokers. Although welders generally did not show serious pulmonary insufficiency, they experienced larger reductions in pulmonary function parameters with increasing age than did the controls.

In 1978, Ross reported the results of respiratory examinations on welders and controls to determine possible health differences due to exposure to welding fumes. Physical examinations were performed on 926 welders between the ages of 20 and 59 and on 755 "other available" non-boilermaker workers. A comparison of welders and controls for respiratory illness as a whole and pneumonia and bronchitis specifically showed that, the only significant difference exists among smoking welders in the 50-59 age group who showed higher percentage of the above three conditions. The only significant difference found between the welders and the control group when tested for FVC, FEV, and PEF was that welders had better mean values than the controls. When tested for restrictive ventilatory capacity, smoking welders aged 40-49 showed a significantly higher prevalence of problems than did controls; conversely, non-smoking controls of the same age group had significantly more cases of restrictive capacity than did welders. Non-smoking welders aged 40-49 were significantly more affected by obstructive ventilatory capacity (i.e., FEV/FVC was below 0.75), whereas the controls aged 50-59 were more affected. Differences between welders and controls were infrequent and did not always favor the control group. Overall, 8.4% of smoking welders and 6.6% of non-smoking welders had restricted ventilatory capacity and 24.4% and 20.3% had some obstruction to their ventilatory capacity, respectively. Also, 7% of the welders had X-ray evidence of siderosis, and

7% had other abnormalities. Those retired welders with over 15 years of exposure had a 30% prevalence of siderosis.

Anthony et al. (1978) reported on a 26-year old man with 9 years of oxyacetylene welding experience who developed a hacking cough soon after having begun a new job. Shortness of breath, a dry cough, and chills began after the first day of soldering brass with a silver solder. Upon analysis, the solder was shown to contain high percentages of zinc (18%) and cadmium (6%). The condition persisted despite medical treatment, and 4 days after exposure, the man had to be taken to an emergency room. While in the hospital, lung function and other tests were performed. Within 3 days the patient's symptoms subsided without treatment. Table 2-1 shows the results of lung function tests on the day after admission and 6 weeks thereafter. Urinalysis by atomic absorption spectroscopy revealed concentrations of 4 mg/dL for zinc and 1.5 mg/dL for cadmium on the tenth day after admission and concentrations of 0.18 mg/dL and 0.53 ug/dL 3 months after exposure. No evidence of airway obstruction was found. X-rays showed pulmonary edema and hemorrage upon admission, but this condition cleared up entirely 6 days later.

Anthony et al. stated that the initial course of this illness, i.e., respiratory difficulties and high urine zinc concentrations, was consistent with metal fume fever. However, metal fume fever is usually self-limiting and lasts no more than 48 hours. The persistence of clinical symptoms for 5 days, elevated urine concentration of cadmium 15 days after exposure, and the finding of substantial concentrations of cadmium in the solder suggested a diagnosis of acute toxic pulmonary reaction to cadmium.

Table 2-1
Results of pulmonary function tests on man with toxic reaction to cadmium

	Observed values	
Variable	Day after admission to hospital	Six weeks later
Total lung capacity, L	3.4	5.5
Functional residual capacity, L	2.4	3.0
Vital capacity (VC), L	1.8	4.1
Residual volume, L	1.7	1.4
Forced expiratory volume in 1 second, L	1.6	3.9
Maximum expiratory flow at 50% VC, L/s	2.7	7.6
Maximum expiratory flow at 50% VC, VC/s	1.5	1.4
Airway resistance, cm H_2O/L/s	1.5	0.9
Diffusing capacity of carbon monoxide, *mL/min/mm Hg	13.7	27.8

*Measured by single-breath method.

2.1.1.5 Radiographic Examination

Attfield and Ross (1978) reported the results of the examination of 661 chest X-rays of electric-arc welders. All welders were currently employed or had retired from welding in a heavy industrial plant. Their ages ranged from 16 to 73 (average of 36.8 years) and posteroanterior radiographs taken between March 1970 and October 1973 were obtained for all men.

Welding dust and fume concentration were not recorded so length of exposure was used as an index of exposure. Because welders of the subject plant generally began their apprenticeship at age 16 and remained in that occupation for life, length of working life was estimated to be present age minus 16 years. The types of electric-arc welding done by these men included GMAW and GTAW processes. Oxides of nitrogen, manganese, nickel, chromium, zinc, iron, and lead were among the contaminants encountered. The vast majority of men welded full-time and approximately 40% of them had, at one time, been exposed to asbestos.

Three physicians with experience in evaluating radiographs read the films; 649 X-rays were read by all three and no one reader read all 661 films. Films were classified on the basis of the amount of small, rounded opacities, according to the 12-point increasing scale of the International Labor Organization(ILO); the scale ranged from 0/ - to 3/4, in increasing severity. Mean scores for each film were calculated and rounded for the judgments of the three readers. Also, the greatest diameter of the predominant opacities were classified as p, q, or r, (less than 1.5 mm, 1.5-3.0, and over 3.0 mm, respectively).

It was determined that, based on mean readings, 7.9% of the film scores were judged to be in category 0/1 or greater. The boundary between 0/0 and 0/1 was taken to be the dividing line between the absence and presence of small opacities. The readers agreed on the classification of 576 films; all of these were of category 0/0. There was over 93% agreement on the presence or absences of radiologic opacities. However, 3 to 12% of each pair of readers agreed on those films categorized as 0/1 or greater. Paired agreement ranged from 60-80% in opacity size; 15-23% were classified as "p," 63-85% as "q" and 0-13% as "r." Approximately 90% of the films read in common were classified identically when examined for abnormalities other than pneumoconioses. Tuberculosis, pleural changes and bronchitis were among the conditions reported.

Attfield and Ross also examined the relationship between opacity prevalence and length of working life. Very few abnormalities were noted until the worker had been employed 15 years, and the prevalence of pneumoconiosis steadily increased thereafter until it passed 30% with 50 years of experience. An analysis of length of experience by severity of the opacities yielded an approximately direct relationship until age 50 when no such increase continued.

Several limitations of this study cloud the applicability of the results. Reader bias may have been introduced by the fact that all readers read the films in the same order and that more than one film was obtained on some workers. It is possible that those welders not in this study, i.e., those who left the industry after suffering adverse effects of welding fumes, may have biased the results. The biggest disadvantage of this study was the use of length of exposure as a measure of fume exposure. Undoubtedly, conditions varied for welders individually and for the occupations as a whole.

2.1.2 Effects on the Ear and Hearing

Jarzebski (1979) reported on 19 cases of middle ear burns, due to sparks or foreign bodies, that occurred in metal welders in Poland during the period of 1970-1976. No discussion was given on the use (or lack of use) of protective devices.

2.1.3 Effects on Eyes and Vision

Welding is associated with five potential sources of hazard to eyes and vision. These sources are fumes, radiation, heat, magnetic fields, and flying sparks. Appropriate protective measures for the eyes and face are available and widely used (NIOSH, 1978, 1979). However, approximately two incidents of eye injury per welder per year have been reported (Tengroth and Vulcan, 1977). Pollak and Romanchuk (1980) maintained that most of the accidents happen to semiprofessional or untrained personnel who do not fully appreciate the hazard of exposure to high intensity radiation and visible light. In the absence of protective measures, welders might experience eye irritation, corneal and conjunctive injuries, arc eye, retinal damage, cataracts, sluggishness of eye adaptation.

2.1.3.1 Eye Irritation

Several factors can contribute to the eye discomfort experienced by unprotected welders. These factors include gases, fumes, and radiation.

Ozone, if present at high levels, might produce eye irritation. However, studies by Bloom (1979) showed that in the respirable zone of welders, ozone never reaches high concentration because of the formation of nitrogen oxides.

Fumes, on the other hand, especially those containing basic materials, have been proven to cause considerable eye irritation. Bases react with lipids in the corneal epithelial cells, forming soluble soap that can quickly penetrate the corneal stroma and enter the anterior chamber of the eye. Acid mists or vapors are generally more tolerable than are basic ones.

There is consideable disagreement on the use of contact lenses by welders.

2.1.3.2 Retinal Damage

Any biological consequence of exposure of the human eye to intense light or optical radiation is caused by thermal and photochemical effects. Thermal damage is caused by visible and infrared radiation, whereas photochemical lesions are due to ultraviolet radiation and are dependent upon photon energy.

The retina is the most sensitive structure in the eye to the thermal effects of visible and infrared radiation. A major part of the absorbed energy is converted to heat. Exposure to a relatively high intensity beam for some time is needed to bring about serious injury to the retina.

On the other hand, exposure to ultraviolet radiation causes very little initial discomfort and might, in severe cases, cause irreversible photochemical reaction in the pigment layer. Permanent damage, however, is mostly caused by thermal effects which might be aggravated by the simultaneous photochemical reaction due to ultraviolet waves.

Pollak and Romanchuk (1980) reported on the use of several mathematical equations for the determination of the thermal effects of optical radiation. The authors applied these equations to the calculation of the temperature of the retina of a postman who had been welding an implement without eye protection. Details of this accident are mentioned by Romanchuk et al. (1978). The postman had been welding a piece of steel without any protection for his eyes except for his regular eyeglasses. The welded object was about 1 foot away from his eyes; and the victim used a class 6013 rod with 220 volt, 95 ampere current. The metal electrodes were 3 mm in diameter.

Calculations made by Pollak and Romanchuk (1980) showed a possible temperature increase of 10.3° C, which was sustained for about 10 seconds. Clinical examination of the patient revealed retinal burns that could not possibly be produced by such a temperature for this short duration. Pollak and Romanchuk maintained that since the final expression of calculating the temperature increase of the retina involved the use of estimated values, the calculated values are only "crude" approximations.

Lovsund et al. (1979 b) studied the effects on vision of a phenomenon called magnetophosphenes, the result of an extremely low frequency (ELF) magnetic field produced during welding. The studies included are on welding [manual metal arc (SMAW) and submerged arc welding] and resistance welding (flash welding, seam welding, and spot welding). The welding processes generated magnetic fields in the range of 10-50 Hz and 0-40 mT. A sensitivity maximum was found at 20-25 Hz. The authors found no evidence of any "alarming" effect on the visual system from short-term exposure to these magnetic fields. However, their results do not permit any definite conclusions on the chronic effects (due to long-term exposure) of welding.

2.1.3.3 Effects on Readaptation Time

The light sensitivity of an eye adapted to a certain luminous level decreases abruptly when the eye is exposed to a bright flash of light. Since welders are usually exposed (especially when proper protective measures are not used) to bright lights, gases, and fumes, the readaptation time (RAT) of their eyes might be affected. The RAT is the time that elapses from the moment of exposure to a bright light until a certain object reappears. Whether this effect on RAT is due to gases or fumes or both and whether RAT changes can be correlated to the content of certain substances in the electrodes was studied by Linde (1980). He reported that professional welders usually experience greatest discomfort when using two types of electrodes: copper- and fluoride-containing rods. This discomfort is partly due to the effects of fumes on the respiratory system and partly due to visual and central nervous system effects. The method of measurement was dependent on the determination of the recovery time for optokinetic nystagmus following a bright flash of light. Welding fumes, produced by different electrodes, were analyzed for particles. Blood samples were analyzed for trace metal contents just before exposure to welding fumes and immediately afterward. A marked increase in RAT was observed with fumes from basic electrodes similar to AWS 7018 and alloyed versions (which contain high amounts of calcium and fluorides and sometimes copper). Fumes from rutile electrodes, on the other hand, produced no marked increase in RAT. These fumes usually contain low levels of calcium and fluorides. The author ascribed the prolongation of RAT to the fluoride content of the fumes. However, blood analysis failed to show any changes that could be correlated with the effect on RAT. Headaches and nausea experienced by welders were found to correlate with the prolongation of RAT.

2.1.3.4 Welding and Contact Lenses

Dixon (1978 b) traced the history of contact lenses and their industrial use, made recommendations, and stated precautions about their use in the working environment. Regarding their use in welding operations, Dixon discounted the story of the welder who, following an arcing episode, "removed part of his cornea as well" when he attempted to remove his contact lenses. No arc injury to the cornea occurred, and it was simply a case of superficial corneal irritation from continuous wearing of the lenses. It was recommended that adequate safety protection in addition to the contact lenses being worn required a qualified person be immediately accessable to remove the lens in an emergency and that the wearer quickly be able to obtain clear vision either with glasses or if his uncorrected vision is adequate.

In a letter to the *British Medical Journal* 1 year earlier, Kersley (1977) presented viewpoints of various professionals on the potential risk of arc welding flash on the eyes of the contact lens wearer. One ophthalmologic surgeon was stated to have favored the use of contact lenses in conjunction with safety glasses in welders exposed to arc flashes due to the added protection they provide. Another surgeon felt that contact lenses offered the wearer adequate protection. Kersley stated that adverse attitudes toward contact lenses are the result of misreporting of a 1967 incident involving a welder who reportedly lost his sight following an arc flash. However, according to the medical director of the involved plant, the damage, which was reversed within a few days of the injury, was caused by neglect of the worker and continuous wearing of lenses.

2.1.4 Effects on the Skin

Most skin conditions in welders are caused by flying hot metal particles and exposure to ultraviolet radiation. Skin conditions reported in welders were always due to lack of protective measures. For instance, Jirasek (1979) reported on ten cases of occupational hyperpigmentation that has the same features as siderosis of the skin. These pigmentation which consisted of disseminated rusty brown minute macules similar to freckles, occurred on the bare arms and forearms of 10 women employees whose work involved spot welding of car parts. No pigmentation were noticed on their hands and wrists which were protected by leather gloves. There were no complaints of any pain or hinderance in work due to this condition. However, it was only a matter of cosmetic significance. The smaller macules disappeared within three years of cessation of exposure, whereas larger macules persisted for as long as 8 years after giving up welding. Contact eczema due to nickel was reported in unprotected electric welders (Weiler, 1979).

No skin cancers attributable to welding were reported. Epidemiological analysis of 200,000 cases of skin cancer conducted by MacDonald (1976) was not occupationally oriented. However, the author attributed the cancer conditions to several factors such as hereditary, exposure to ultraviolet radiation, latent virus etc. There is a potential exposure of welders to ultraviolet radiations; however, the contribution of this exposure to the incidence of skin cancer is unknown.

2.1.5 Effects on the Gastrointestinal Tract

No article was published in the 1979-80 period of updating the previous report on welding.

2.1.6 Effects on the Cardiovascular System

Various cardiovascular myopathies have been ascribed to exposure to welding fumes (Abramovich-Poliakov et al., 1979; Suvorov et al., 1980). No details were given.

2.1.7 Effects on the Central Nervous System

According to our best knowledge, no article on this subject has been published in the 1979-80 period.

2.1.8 Effects on the Liver

No article on this subject could be located in the 1979-80 period.

2.1.9 Effect on the Musculoskeletal System

An epidemiologic study on the effect of welding on the knee joint was conducted by Nauwald in 1980 on 100 welders working in the shipbuilding industry. Although the number of workers is too small to allow any significant conclusions, the author claimed that welders are at an increased risk of developing knee joint diseases. The study was stratified by age, and bursa changes were found in all age groups, especially the 46 - 55 group. No details were given on previous occupation or the factors that predisposed to the knee joint ailments.

2.1.10 Effects on Reproductive System

No article dealing with this subject has been published in the 1979-80 period.

2.1.11 Effects on the Urinary System

No article on this subject has been published during this period of updating (1979-1980).

2.1.12 Effects on the Endocrine System

No articles on this subject could be located.

2.1.13 Effects on the Teeth and Oral Cavity

No article dealing with these effects has been published.

2.1.14 Metal Fume Fever and Allergic Reactions

Calnan (1979) described the condition of a 42-year-old man who had a 4-year history of a recurring rash (diagnosed as endogenous eczema) on his left leg and the backs of his hands. The man had been a welder in a small shipyard for about 10 years. He had welded both galvanized water pipes and mild steel, which occasionally was treated with red lead. Welding galvanized metal was reported to have generated zinc fumes, as the surface zinc vaporized. Over the 2 years prior to examination, he had had and quickly recovered from, malaria-like attacks of a "shivering illness." Calnan described the illness as typical of metal fume fever.

2.1.15 Biochemical Effects of Heavy Metals in Welding Fumes

The term "heavy metals" denotes those metals that have a density greater than 5 gm/cm^3. It covers about 40 elements, some of which are highly toxic. The toxicity of heavy metals in general was discussed by Louria et al. (1972) and Clarkson (1977).

Since fumes generated by welding contain particles of various metal oxides, several authors have tried to determine the extent of exposure to welding fumes by determining the level of metals in different biological specimens. Thus, Grund (1980) and Grund et al. (1980) analyzed the manganese contents of hair samples obtained from 254 shipyard welders and 237 control persons by neutron activation method. The manganese content in welders' hair was five times as high as that of the control group. Furthermore, correlation was found between welding technique and the amount of contamination. However, no correlation could be established between the serum and hair contents of manganese. Since manganese is preferentially stored in bradytropic tissues such as hair, it seems that hair analysis, which is a non-invasive test, would be a suitable test for early identification of exposure to manganese. The major drawbacks of this test are that (1) no correlation could be established between manganese content and years of exposure or age and (2) that no correlation between manganese content in the hair and lung function data could be found (Grund et al., 1980).

Bernacki et al. (1978) analyzed the nickel concentration in the urine of nickel-exposed workers and unexposed controls. Using electrothermal atomic absorption spectrometry, urinalyses were performed on 10 groups (101 total workers) exposed to nickel (including a group of 10 arc welders) and two groups (42 workers) of unexposed controls. When the mean urine nickel concentrations of the 10 arc welders (7 men and 3 women) were compared with those of a control group of 19 hospital workers, it was found that the calculated nickel concentration of 6.3 ug/1 of urine for welders was significantly greater than that of controls. The nickel concentrations per gram of creatinine were not significantly different. It was claimed that measurement of urine concentrations of nickel are valuable assessments of workplace exposure and are more sensitive and convenient procedures, even though no significant correlation was observed for the welders between atmospheric concentration of nickel in the workplace and its concentration in the urine.

Tola and Karskela (1976) reported the results of a survey of lead exposure in 568 Finnish workers, including 84 welders, in three shipyards. Using Hessel's method of determining blood lead concentrations, it was found that none of the shipyard workers had blood lead concentrations above the Finnish highest recommended concentration of 70 ug/100 ml. The mean blood lead concentration of the welders was approximately 26 ug/100 ml. The hemoglobin levels of all workers studied fell within the normal range, and no statistically significant differences between any subgroups were noted. It was concluded that these low lead concentrations indicated a low risk of lead poisoning in the workers studied.

Bergert and Voigt (1980) and Mutti et al. (1979) estimated the urinary content of fluoride and chromium, respectively, in welders. Experiments performed by Bergert and Voigt (1980) utilized 8 persons; each was given 1-2 mg of fluoride orally. About 10% of administered fluoride was excreted in the urine in 4 hours. The authors concluded that the estimation of fluoride content in urine of welders, especially those using potassium-containing electrodes, may act as an indicator of the respiratory load of welding fumes. It should be pointed out that in Bergert and Voigt's experiment, the number of persons used is inadequate to draw any meaningful conclusion. Furthermore, the conditions of the experiment did not simulate the real atmosphere of welding, since fluorides were given orally; inhalation route would have been a closer approximation of the actual exposure.

Experiments performed by Mutti et al. (1979) were designed to measure the chromium content in urine of welders during the actual exposure to welding fumes. Twenty-two welders using high chromium alloyed electrodes were examined. Measurements were made of the hexavalent hydrosoluble fraction of chromium in the welding environment and also in the urine of the welders. A correlation between the amount of airborne and urinary chromium was found. Urinary excretion of chromium was higher in individuals with greater chromium body burden. These findings led the authors to conclude that if the urinary excretion of chromium is not significantly increased after a single working period, then it could be assumed that significant absorption of the hexavalent chromium did not occur.

The correlation between the atmospheric level of chromium and nickel and their biological half-life of urinary excretion was further investigated by Tossavainen et al. (1980). It was found that a linear one-compartment model gave estimates of half-life ranging from 15 to 41 hours for chromium in urine and 17 to 39 hours for nickel.

Mlynarczyk and Senczuk (1980) claimed that the decrease in plasma proline and the concomitant increase in plasma hydroxyproline could be used as an early indication of exposure to welding fumes.

Wysocki et al. (1980) studied the effect of welding on serum enzyme levels in two groups of welders who averaged 3.5 and 7.2 years of experience, respectively. They found a significant decrease in the alkaline phosphatase level in both groups and a significant decrease in the lactate dehydrogenase enzyme.

2.2 Mutagenicity and Carcinogenicity of the Fumes and Gases

Because of the intense ionization of the air during welding, ozone is formed to a variable degree. There is growing evidence that ozone induces both point and chromosomal mutations. A concentration of 0.2 ppm of ozone brought about an increase in chromosomal aberrations in Chinese hamsters lymphocytes (Zelac et al., 1971). Chromosomal aberrations due to exposure to ozone have been reported in chick embryo fibrotolsis (Sachsenamier et al., 1965) and in Vicia fava (Fetner 1962). Point mutation was demonstrated by Prat et al. (1968) who found that ozone alters the pyrimidine bases thymine and uracil in E. coli DNA.

These findings have prompted an *in vivo* study of the effects of ozone on human chromosome response (Bloom, 1979). A total of 247 welding trainees were involved in Bloom's study, each serving as his own control. Blood samples were taken prior to ozone exposure, and 6 and 12 weeks after the start of a 12-week training period. Seventy-four percent of the trainees were between the ages of 18 and 21 years, and almost half of the trainees had had previous school or job welding experience. Only 131 trainees completed the program and thus provided the third blood sample. The maximum ozone level detected during welding was 0.015 ppm both in the ambient air of the welding sheds and in the respiratory zone within individual welding booths. A slight rise in the percentage of cells with single chromatid and isochromatid gaps was demonstrated in the 6 week samples, which declined by the twelfth week. However, no statistically significant increases were seen in any category of the chromosome-type aberrations studied. The slight increases in the single and isochromatid gaps do not indicate any permanent damage and are well known to be a less reliable indicator of chromosomal damage than any other aberration type. The strikingly low level of ozone detected appears to be due to the rapid conversion of ozone to the oxides of nitrogren.

An *in vitro* study was conducted by White et al. (1979) to study the effects of components, other than ozone, generated during welding, namely fumes. Because hexavalent chromium may be linked to the increased incidence of cancer in workers from chromium-associated industries (Langard and Norseth, 1975; Royle, 1975), including the welding industry (Milham, 1976), White used an established human cell line NHIK 3025 (originating from cervix carcinoma) to study the possible effects of welding fumes. Fumes were obtained by welding stainless steel plates (23.5% chromium and 21.5% nickel) with a stainless steel electrode (18% chromium and 11% nickel) at 90 amperes in a specially constructed chamber; fume particles were collected on an Acropore AN 1200 filter (mean pore size 0.2 um). Fume particles were then suspended in normal saline and were added to the culture medium of the cells. The addition of fume particles resulted in a reduced cell proliferation and substantially increased the fraction of abnormally large cells. The soluble part of fumes produced the same effect, and it was found that hexavalent chromium constituted 3.5% of the total welding fume material. The insoluble particle fraction showed no activity. It was concluded that the hexavalent chromium content of the welding fume particulates is probably the main cause of the toxic effects observed with NHIK 3025 cells.

2.3 Epidemiologic Studies

Beaumont (1980) studied the mortality of welders, shipfitters, and other metal trade workers in the Greater Seattle area. That study, though a sincere effort, could not reach definitive conclusions on the cause and effect relationship of various diseases in welders because of different uncontrolled factors. For example, the Boilermakers Union in the Seattle local includes workers who perform 16 different jobs; these workers usually perform more than one type of work during each year. Beaumont exemplified this heterogenicity by stating that "one welder, will often work for a shipyard, a metal fabrication shop, and a field construction job, all in the same year." Furthermore, in 1958, the Welders and Boilermakers Unions merged, and records from the merged unions were pulled together and used to represent welders. Exposure to hazardous substances such as asbestos, as well as smoking, might have been contributing factors in the occurrence of diseases in welders. In fact, Beaumont stated that most of the welders might have been exposed to asbestos in shipyards where they worked at one time or another. He also maintained that welders smoke cigarettes more than other occupational groups and the nation as a whole. Beaumont's results showed that cancer of the respiratory system was higher than expected (31%) in welders; however, this increase was not statistically significant. The lack of detailed exposure data on individual workers did not allow the association of the increased risk with any particular substance. Beaumont found that persons engaged in welding for 20 years or more are at higher risk of lung cancer than others.

Concerning pneumonia in welders, Beaumont found a 67% excess of pneumonia death (19 observed and 11.4 expected). The author associated the occurrence of pneumonia with the "considerable amounts" of nitrogen dioxide and ozone involved during welding.

Exposure of welders to ultraviolet radiation might affect the lens of the eye, causing the so-called "arc eye," and can also cause skin burns. In this epidemiologic study, mortality from skin and eye cancers among welders was not unusual (2 observed and 2.1 expected). Beaumont also found that the occurrence of prostatic carcinoma among welders was in the normal range.

Preliminary results of an epidemiologic study conducted by McMillan (1979 b,c; 1980) on the health of welders in the Royal Dockyards have been presented to the British Occupational Hygiene Society. Although the study took into account a series of factors such as morbidity, mortality, and clinical studies on chronic respiratory disease, data analysis was incomplete and results may not represent the final findings.

The preliminary findings of McMillan showed that British welders included a high percentage of smokers; there was no evidence of a significant excess of chronic respiratory diseases. A high rate of musculoskeletal disability in welders was also noticed. This was attributed to the fact that welders comprised a high proportion of older men; whether this was due to welding or simply aging was not clear.

3. Toxicologic Investigations in Animals

Experiments performed on various animal species are an important source of our knowledge of the toxic effects of different chemicals. A well-designed experimental study on animals should simulate, as much as possible, the normal atmosphere of exposed workers, as well as the route of entrance into the body. Therefore, conclusions derived from experiments in which welding fumes are administered, say, intraperitoneally, are questionable.

To assess any potential health effect of welding, experiments have been focused on fumes and radiation generated during welding. Several animal species have been exposed to welding fumes, e.g., rats (Hewitt and Hicks, 1972; Samoilova and Kireev, 1975; Guskova and Komovnikov, 1974; Arutyunov et al., 1976), guinea pigs (Kawada et al., 1964), mice (Von Haam and Groom, 1941), cats (Titus et al., 1935), and rabbits (McCord et al., 1941; Garnuszewski and Dobrynski, 1966).

These previous studies were mainly concerned with the local effects (such as irritation and pulmonary edema) of welding fumes on the lungs; recent investigations, however, have focused on the pattern of deposition and the elimination rate of fume particles from lungs (Lam et al., 1979; Alshamma et al., 1979a,b).

In all these studies, fumes were generated by welding various metals using different welding methods. Animals were exposed to fumes by inhalation or received certain amounts of the fumes by intratracheal instillation. Although these experiments are pivotal in our understanding of the effects of welding on health, data should be interpreted with extreme caution for the following reasons:

1. The diameter of the respiratory passages of man is different, usually larger, than those of experimental animals. This will affect air velocity and hence the deposition pattern of particulates. The site of deposition, in turn, determines the degree of absorption of compounds leached from particles and hence their systemic toxicity, the severity of local effects, and the clearance mechanisms of these particles (Doull et al., 1980).
2. The respiratory rate in man (about 12/minute) is different from that in other experimental animals e.g., 100/minute in rats).
3. Intratracheal administration of welding fumes in experimental animals will result in the introduction of a much wider range of particle sizes and a much deeper penetration by particles in lungs than would normally occur by inhalation.
4. In some experiments, total body exposure of animals was performed, while in others, nose only exposure was carried out. Both techniques have their limitations. Deposition of fume particles in the fur of animals exposed by the former method might change the toxicity profile and the outcome of the study. In the latter case, viz., when animals are restrained in a confined space to have only the nose exposed, exposure might result in a change in respiratory rate and pattern. An animal holding its breath could change the amount and locus of fume deposited in lungs.
5. Fume concentrations in animal experiments are usually much higher than those normally encountered by welders. This may result in "poisoning" or saturation of the elimination channels and hence could give a totally misleading pattern of excretion of fume particles.

An attempt to study the deposition and elimination of inhaled particles of welding fumes was carried out by Lam et al. (1978). Because it highlights some of the above-mentioned reservations, this study will be discussed in detail. Experiments were performed on Sprague-Dawley male albino rats and Dunkin-Hartley albino guinea pigs. The latter were used because guinea pigs are appropriate for hypersensitivity studies; it was suspected from elemental analysis that GMAW fumes could induce hypersensitivity. Guinea pigs were exposed to a concentration of 990 mg/m^3 of welding fumes produced by the metal-inert gas process (GMAW) using stainless steel wire. Exposure continued for 256 minutes when animals displayed signs of respiratory distress; in fact, two animals died soon after. The fractional deposition (the ratio of the amount of fume deposited in animal lungs to the total amount generated) of GMAW welding fumes in guinea pigs was 0.17.

Rats, on the other hand, were exposed to 1,178 mg/m^3 of fumes generated by the manual metal arc (SMAW) process using flux-coated electrodes for 45.8 minutes. Exposure was terminated because of "undue rise of atmospheric temperature." The average lung deposit was 1.5 mg*, and the fractional deposition was 0.23. The particle size and composition of the fume were as follows:

	GMAW (guinea pig)	SMAW (rat)
Particle size (count median diameter)	0.064 um	0.125 um
Composition (ug/g)		
Antimony	22	-
Chromium	110,000	940
Cobalt	170	89
Iron	609,000	319,500
Nickel	56,000	-
Zinc	1,500	660

The amount of the fume deposited in the lungs was determined immediately after the termination of exposure and at intervals up to 30 days in the rat study and 80 days in the guinea pig study. A marked decrease by time in the lung concentrations of all elements studied was observed. The concentration of iron in the lung tissues of rats decreased from an initial level of 2,185 ug/g of lung tissue immediately after the exposure to about 1,085 ug/g (50%) 30 hours later. A concentration of 657 ug/g was found 30 days later. The figures for cobalt, in the same intervals, were 0.28, 0.12, and 0.09 ug/g, respectively. Those of chromium were 32, 10, and 2, respectively. Overall, iron was retained longer than chromium or cobalt. Similar decreases were obtained with guinea pigs. However, unlike the rat experiments, about 40 days were needed for the elimination of 50% of the iron content of guinea pigs lungs. The results of various elements were as follows:

Time after exposure	Fe	Co	Ni	Cr
	Mean value (ug/g) of freeze dried lung			
0 hour	10,563	17.1	135	3928
10 hours	9,008	13.6	101	3118
5 days	6,792	10.9	81	2288
40 days	5,207	7.2	52	1367
80 days	4,441	5.0	35	908

*On the basis of the relative surface area of man and rat, a deposit of 1.5 mg in rat lungs is equivalent to 83 mg in human lungs. Kalliomaki et al. (1980) measured the amount of dust in 34 welders' lungs and found an average of 139 mg after 9.9 years of welding. The duration of exposure in Lam's experiment was 45.8 minutes.

To determine the biological fate of the fumes once they entered the lungs, Lam et al. (1978) injected 10 mg of radioactive GMAW fume particles (total radioactivity used was 1 u Ci) into the trachea of anesthetized guinea pigs. The elimination pattern was similar to those of inhalation studies; that is, chromium was eliminated quickly and cobalt and iron less rapidly. Three phases were distinguishable in the elimination process of welding fumes. Phase I is characterized by a short half-life and was ascribed to the elimination of particulates by the mucociliary escalator. In this phase particles were transported in their entirety without any leaching of various constituents. Phase II, which is slower than Phase I, accounts for longer half-life constants and is due mainly to phagocytosis and transport of particulates by macrophages. Phase III is much slower, has a biological half-life of several weeks, and is achieved mainly by leaching. The biological half-life periods of the three phases for different metals were as follows:

Biological half-life (days) of radioactive GMAW fumes

	Biological half-life		
	Phase I	Phase II	Phase III
59Fe	0.61	6.5	318
60Co	0.81	8.3	229
51Cr	0.68	5.6	135

It is noteworthy that iron was found only in the feces of animals, whereas chromium and cobalt were detected in both feces and urine.

Ideally, radioactive fumes should be inhaled by animals. However, intratracheal administration was performed by Lam et al. (1978) and Al-Shamma et al. (1979a) for safety reasons, and to be sure of the dose administered. A dose of 10 to 15 mg of neutron activated GMAW welding fumes was instilled intratracheally in guinea pigs in the Al-Shamma study. Fourteen days later, the highest concentration of iron, cobalt, and chromium was found in lungs, followed by liver and kidneys. Radioactivity was also found in other organs such as spleen, heart, testis, brain, and pancreas.

Experiments performed by the same authors on rats Al-Shamma et al. (1979b) used a more reasonable concentration of welding fumes (380-400 mg/m^3). Male Sprague-Dawley rats were exposed to MIG fumes for a continous period of 184 minutes. Animals were then sacrificed at different time intervals up to 7 days post exposure. Another group of rats was exposed to GMAW fumes for 30 minutes/day for 6 days. After a period of 40 days, animals were again exposed to the fumes for 30 minutes/day for 8 days. The total exposure time was 431 minutes. Significantly high levels of iron, cobalt, and chromium were found in the lung tissues of exposed animals. The concentration of these elements progressively and rapidly decreased over a period of 7 days post-exposure. An increase in the spleen content of iron and cobalt was observed in the exposed animals as early as 3 hours after cessation of exposure. This indicates a rapid translocation of the soluble fractions of these elements. Most importantly, the authors found that the rate of elimination of these elements from the lung is slower following the intermittent inhalation than in single exposure. This phenomenon might be caused by the cumulative deposition of these elements, or due to interference with the pulmonary clearance mechanisms.

4. In Vitro Studies

In view of the fact that cancer usually develops in humans after a latent period of 5-40 years, several "short-term" screening techniques have been developed to test for potential chemical carcinogens. Among these tests are those for mutagenesis, or the ability to induce permanent change in genetic material in bacteria. These procedures depend upon the assumption that cancer is related to genetic alterations; however, this assumption was not true for some chemicals. The value of the short term tests, therefore, is to raise a red flag when results, especially from more than one species, are positive.

Mutagenicity of fumes from different welding operations has been studied in bacteria (Hedenstedt et al., 1977; Maxild et al. 1978) and culture cells (Hedenstedt et al., 1977). Both Stern (1977, 1978) and Maxild et al. (1978) found that the water-soluble portions of fumes obtained by welding stainless steel are mutagenic; those obtained from mild steel are not. The mutagenic effect in these *in vitro* systems was found by Maxild et al. (1978) to be particularly associated with particles produced by welding stainless steel using the manual metal arc (SMAW) method. Those obtained using the metal inert gas (GMAW) method were less mutagenic. The difference in mutagenicity was attributed to the particle size and water solubility of these fumes.

Cytogenic effects of welding fumes were further studied in 1979 by Koshi. The frequencies of sister chromatid exchange (SCE) and chromosome aberrations were examined in culture Chinese hamster cells exposed to welding fumes collected from SMA and GMA welding. Particles from SMA welding were 100 times more mutagenic than those from GMA welding. Fumes from SMA welding contained 60 times more soluble hexavalent chromium than those from GMA welding. It was assumed, therefore, that hexavalent chromium may be involved in the mutagenic effects of the fume particles from stainless steel welding. This supports the conclusion derived earlier by Hedenstedt et al. (1977).

In 1980, Knudsen used the so-called mammalian spot test to study the mutagenic effects of welding fume particles. This test, which depends on the development of colored spots in the fur of the offspring 2-5 weeks after pregnant female mice are treated with a given compound, is in an early stage of validation as a useful tool for the detection of mutagenicity. Knudsen found that intraperitoneal injection of pregnant female mice with 100 mg/kg of fume particles, collected from SMA welding of stainless steel, resulted in the development of grayish or brownish spots in the fur of offspring. Apart from the questionable validity of this test, the dose seems to be very large, and toxicity might be due to a host of effects in both the dams and the offspring. Furthermore, intraperitoneal injection of welding fumes does not represent the inhalation route by which welders are exposed to fumes. So, the relevance of this study to the effect of welding on health is questionable.

In addition to the fact that not every mutagen is a carcinogen, mutagenicity of filter-collected mutagens could be artifactual (Chrisp and Fisher, 1980). Because of the potential for chemical reaction of the collected particles, great caution should be exercised in interpreting these results. For instance, perylene, a non-mutagenic compound in the Ames test, was converted to the mutagenic derivative 3-nitroperylene after exposure to nitrogen dioxide on fiberglass filters.

5. Conclusions and Recommendations

Protection of welders from the adverse health effects of fumes and gases generated during welding is a major concern of the American Welding Society (AWS). For this reason, the AWS has entrusted the Franklin Research Center to review the world literature on welding and its effects on health. Two previous reports have been submitted, and the present report is concerned with the literature published from June 1979 to December 1980. The health of welders was the major topic at recent AWS conferences (Milwaukee 1979, and Little Rock, 1980).

Because of the difference between the susceptibility of various individuals to the effects of compounds in the welding environment, McMillan (1979, 1980) emphasized the importance of medical surveillance to detect any ill effects in welders, as well as a pre-employment examination for the welding trainees.

Despite the vast amount of published literature, a host of questions are still unanswered. Research conducted in these "gray" areas will aid in the assessment of health hazards, if any, to welders. The following topics for research are recommended, based on the gaps that have been identified.

1. The threshold limit values for several compounds that are present in the welding environment have been established based on experiments that were conducted several decades ago. Because of the different standards of todays toxicology, as well as the varying sensitivity of individuals, the allowable limits of all compounds present in welding fumes should be scrutnized.
2. Epidemiological studies conducted to date have unfortunately added to rather than helped to clear up the confusion. The problem stems from the fact that no two groups of welders are exactly comparable; even an individual welder could use various electrodes or methods of welding that are inconsistent during his work-life. Social habits such as smoking, drinking, or exposure to other chemicals such as pesticides, insecticides, paint removers, etc., might introduce a new factor that could change the outcome of the study. Interpretation of epidemiological studies should therefore be done with great caution. It would be essential in any future epidemiological studies to account for factors such as smoking, previous exposure to asbestos, previous type of work, familial history of diseases, genetic factors, etc.
3. The lung function test is not a very sensitive method for the detection of any early changes in the respiratory system due to exposure to welding fumes. Methods recently developed, such as magnetopneumography, for the determination of the amount of fume deposited in the lung could be a more accurate measure for determining the extent of exposure. Furthermore, the lung function test should be standardized if it is to serve as an important test.
4. What happens to the fume particles after deposition in the lungs is still unclear. The fate of inhaled particles should be studied in well-designed pharmacokinetic studies that take into account the complexity of the fume composition and the possibility that one component in the fume might affect the fate of another.
5. The alleged presence of chemical carcinogens in welding fumes is a highly controversial issue and needs to be settled. The development of "short-term" tests for carcinogenicity has reduced the time and effort necessary to detect carcinogenic agents. It is recommended that at least three short-term tests be performed in fumes from welding different metals using various operations. It should be pointed out, however, that positive results do not necessarily mean that the fumes are carcinogenic. It would, nonetheless, raise a "red flag" for more testing.

6. A pre-employment screening for health problems could be helpful in eliminating persons with inherent diseases that would be erroneously ascribed to welding. Periodical medical examination and administration of medical questionnaires are advisable.
7. Efforts should be continued toward reduction of the exposure of all those in the welding environment through
   a. improving industrial hygiene practices
   b. use of processes or consumables that produce less fumes
   c. improving ventilation practices
8. The use of protective measures and more extensive safety training of welders and supervisors, and more aggressive supervision and monitoring cannot be emphasized enough. The majority of ill effects were reported when welders neglected the use of safety devices.
9. Development of new methods for the analysis of trace amounts of metals in air and in biological fluids could be helpful.
10. The use of non-invasive techniques such as the use of the metal content in hair or nail to assess exposure is still in its infancy. Further validation of these techniques is necessary.

References

Abramovich-Poliakov DK, Kleiner AI, Kolodub FA, Krakovskaia SP, Nedbailo EP: Clinical features of the effect of magnetic fields in contact electric welding, *Vrach Delo* 4:106-10 (1979).

Akbarkhanzadeh F: Effect of welding variables on the fumes and gases in the welder's breathing zone. *Metal Construction* 11:503-05 (1979).

Akbarkhanzadeh F: Long-term effects of welding fumes upon respiratory symptoms and pulmonary function. *J. Occup. Med.* 22:337-341 (1980).

Akselsson KR, Desaedeleer GG, Johansson TB, Winchester JW: Particle size distribution and human respiratory deposition of trace metals in indoor work environments. *Ann. Occup. Hyg.* 19:225-38 (1976).

Al-Shamma KJ, Hewitt PJ, Hicks R: Distribution of some metallic components of inhaled MIG welding fume in the rat. *Proc. Int. Symp. Nucl. Act. Tech. Life Sci.* 657-66 (1979 a).

Al-Shamma KJ, Hewitt PJ, Hicks R: The elimination and distribution of components of welding fumes from lung deposits in the guinea pig. *Ann. Occup. Hyg.* 22:33-41 (1979 b).

American Welding Society: *Fumes and Gases in the welding environment.* 1-232 pp (1979).

American Welding Society, AWS Project Committee on Fumes and Gases: *Laboratory Method for Measuring Fume Generation Rates and Total Fume Emission of Welding and Allied Processes*, AWS FI.2-79. 7 pp (1979).

American Welding Society Subcommittee on Definitions: Classifications of the Welding Process. *Weld Journal* 55: 270-72 (1976).

Anonymous: Electronic air cleaning units overcome flux cored welding fume problem. *Welding Journal* 58:41-42 (1979 a).

Anonymous: Welding industry meeting on the health of the welder. *Welding Journal* 58:61-62 (1979).

Anonymous: Electrostatic precipitation for welding fume/smoke removal involves three filtration techniques. *Welding Jornal* 59:41-42 (1980).

Anthony JS, Zamel N, Aberman A: Abnormalities in pulmonary function after brief exposure to toxic metal fumes. *CMA Journal* 119:586-588 (1978).

Antti-Poika M. Hassi J, Pyy L: Respiratory diseases in arc welders. *Int. Arch. Occup. Environ. Health* 40:225-30 (1977).

Arutyunov VD, Batsura YD, Kireev VI, Likachev YP: Development of Pulmonary Alveolar Proteinosis Following Inhalation of some Industrial Aerosols. *Gig. Tr. Prof. Zabol* (9):33-37 (1976).

Astrop A: Clearing the air on fume control. *Machinery and Production Engineering* 137:34-39 (1980).

Attfield MD, Ross DS: Radiological abnormalities in electric-arc welders. *Br. J. Ind. Med.* 35:117-22 (1978).

Bartley DL, McKinnery WN Jr., Wiegand KR: Control of Physical Agent Hazards of Arc Welding and Cutting. *National Institute for Occupational Safety and Health* National Technical Information Service PB80-164866:50pp (1979).

Beaumont JJ: Mortality of Welders, Shipfitters, and Other Metal Trades Workers in Boilermakers' Local No. 104, AFL-CIO. Thesis. Seattle, University of Washington, 158 pp (1980).

Bement LJ: Totally confined explosive welding. *US Patent* 4 106, 687, 55 pp, Aug. 15, (1978).

Bergert KD, Voigt H: Urinary fluoride excretion as proof of exposure in welding with lime electrodes. *Z. Gesamte. Inn. Med.* 35: 188-189 (1980).

Bernacki EJ, Parsons GE, Roy BR, Mikac-Devic M, Kennedy CD, Sunderman FW: Urine nickel concentrations in nickel-exposed workers. *Ann. Clin. Lab. Science.* 8:184-189 (1978).

Betz GM: Interpreting welding, cutting, and brazing regulations *Plant Engineering* 115-19 (1978).

Betz GM: How OSHA welding standards are interpreted. *Welding Design & Fabrication* 52:119-30 (1979).

Birchfield JR: This shipyard kicked the smoking habit. *Welding Design and Fabrication* 53:134-37 (1980 a).

Birchfield JR: Weld shop ventilation takes thought and care. *Welding Design and Fabrication* 53:94-102 (1980 b).

Bloom AD, Calafiore DC: Chromosomal Abnormalities Among Welder Trainees, EPA-600/1-79-011. Research Triangle Park NC, US Environmental Protection Agency, Health Effects Research Laboratory, 39 pp (1979).

Boekholt R: Workplace environment and productivity. *Jaargang* 43:38-46 (1977 a).

Boekholt R: Welding workplace environment (ergonometrics) and humanizing work. *De Veiligheid* 53:175-81 (1977 b).

Bohgard M, Jangida BL, Akselsson KR: An analytical procedure for determining chromium in samples of airborne dust. *Ann. Occup. Hyg.* 22:241-51 (1979).

Breslin AJ, Harris WB: Use of Thoriated Tungsten Electrodes in Inert Gas Shielded Arc Welding - Investigation of Potential Hazard. *Ind. Hyg. Q* 13:191 (1952).

British Patent, Compagnie Maritime D'Expertises-Comex SA, Comex SA: Method and apparatus for welding underwater. *British Patent* 1 552 340. 1-6, 1976.

Buist AS, Van Fleet DL, and Ross BB: *Am. Rev. Resp. Dis. 107-735* (1972).

Buckup H: The Welder's Lung: Harmful or Indifferent Pulmonary Changes. *Z Arbeitsmed Arbeitsschutz* 23(9):286-93 (1973).

Calnan CD: Metal fume fever. *Contact Dermatitis* 5:125 (1978).

Campbell, DL: Report on tests of welding filter plates, NIOSH Tehnical Information, HEW Publication No. (NIOSH) 76-198. Morgantown WV; US Dept. of Health, Education, and Welfare; Public Health Service; Center for Disease Control; National Institute for Occupational Safety and Health; Appalachian Laboratory of Occupational Safety and Health 29 pp (1979).

Candarella R, Cascella D, Tabaroni G, Corso T, Raffi GB: Occupational diseases of welders: Note 1, Respiratory effects. *Giornale Hal Med Lav* 1:31-37 (1979).

Carta P, Sanna-Randaccio F: Experimental emphysema. Bull. Europ. Physiopath. Resp. 13:561-90 (1977).

Chrisp CE, Fisher GL: Mutagenicity of airborne particles. *Mutat. Res.* 76:143-164 (1980).

Clapp DE, Owen RJ: An Investigation of Potential Health Hazards of Arc Welding Fume Growth with Time. *Welding Journal* 56 (12):380s-85s (1977).

Clarkson TW: Factors involved in heavy metal poisoning. *Fed. Proc.* 36:1634-1639 (1977).

Delong WT: Stainless steel - new information on hexavalent chromium. *Welding Journal* 58:70, 72, 74-75 (1979).

Dixon RD: Measurements of noise on the beam of an electron beam welding unit. *Welding Journal (Research Supple.)* 57:51s-56s (1978 a).

Dixon WS: Contact Lenses: Do they belong in the workplace? *Occupational Health and Safety* 133:36-39 (1978 b).

Doull J, Klaassen CD, and Amdur MO: Toxicology, The Basic Science of Poisons. McMillan Publishing Co., Inc. New York, 2nd Edition, 1980.

Eaton NF: Welding Processes and Their Potential Hazards (manuscript) Canadian Welding Development Institute, 19 pp (1977).

Emmett EA, Horstman SW: Factors Influencing the Output of Ultraviolet Radiation During Welding. *J. Occup. Med.* 18(1):41-44 (1976).

Evans MJ, Ingle J, Molyneux MK, Sharp GT, Swain J: An occupational hygiene study of a controlled welding task using a general purpose rutile electrode. *Ann. Occup. Hyg.* 22:1-17 (1979).

Fetner RH: Ozone-induced chromosome breakage in human cell culture. Nature (London) 194:793-794 (1962).

Frew JW: Collection of steel particles from fume, United Kingdom Atomic Energy Authority, *UK Patient* Application GB2 022 465A (1979).

Funahashi A, Schlueter DP, Siegesmund KA, Pintar K: Energy dispersive X-ray analysis of welders' pneumoconiosis. *Amer. Rev. Resp. Dis.* 119:215 (1979).

Garnuszewski Z, Dobrzynski W: Histological Appearance of the Lungs in Experimental Animals Exposed to Welding Fumes. *Gruzlica* 34:4-14 (1966).

Gibson GJ, Seitz RO: Thoriated-Tungsten Electrodes - Their Welding characteristics and Applications *Welding Journal* 31:102 (1952).

Gobbato F, Gambaretto GP, Rausa G: Pollution caused by manual arc welding—quantity and quality variations of welding fumes as a function of electrode characteristics (diameter, weight, and coat composition). *Ig. Mod.* 72:381-98 (1979).

Gola A, Galazka Z, Tychowski K: Serum alpha-1-antitrypsine and respiratory system status in arc welders. *Med. Pr.* 31:41-44 (1980).

Gonzales ER, Gomez JA: Welding in shipbuilding (La soldadura en la construccion naval). *Medicina y Seguridad del Trabajo* 27:23-28 (1979).

Gonzales J, Young RF, Amato MA: Cored Type Welding Electrodes, Lincoln Electric Co *UK Patent* Application GB2020591 6 pp (1979).

Green GM, Jakab GJ, Low RB, Davis GS: Defense mechanisms of the respiratory membrane. *Am. Rev. Resp. Dis.* 115:479-514 (1977).

Grund W: Incorporation of manganese in human hair, A criterion for assessment of exposure. *Arch. Exp. Vet. Med.* 34:347-351 (1980).

Grund W, Schneider WD, Wiesener W: Manganese content of the hair, A criterion for evaluating exposure risks for electric welders. *J. Radioanal. Chem.* 58:319-326 (1980).

Guskova NI, Komovnikov GS: Effect of Welding Rod Coating Dust on Alveolar Phagocytosis and Collagen Content in the Lungs. *Gig. Tr. Prof. Zabol* 6, 12-15 (1974).

Hagopian JH: Validation of a Recommended Approach to Recirculation of Industrial Exhaust Air. Volume I. Spring Grinding, Chrome Plating, Dry Cleaning, Welding, and Vapor Degreasing Opertions, DHEW (NIOSH) 79-143a. *Govt. Reports Announcements & Index* (PB80-161474 80:2166 (1980).

Head IW, Silk SJ: Integral fume extraction in MIG/CO2 welding. *Metal Construction* 11:636-38 (1979).

Hedenstedt A, Jenssen D, Lidesten BM, Ramel C, Rannug U, Stern RM: Mutagenicity of fume particles from stainless steel welding. *Scand. J. Work Environ. Health* 3:203-11 (1977).

Heiple CR, Dixon RD: (Welding of) Beryllium: Problems of cracking and of environment. *Welding and Metal Fabrication* 47:309-10,312-14,316 (1979).

Hewitt PJ, Hicks R: The Retention of Some Metallic Elements in Tissues of Rats Exposed to Welding Fumes. *Br. J. Pharmacol.* 46(3):571P-72P (1972).

Hinrichs JF: Project committee on radiation—summary report. *Welding Journal* 62-65 (1978).

Honavar DS: Indian welding electrode industry-past, present and future. *Indian Welding Journal* 11:63-78 (1979).

Horstman S, Ingram JW: A critical evaluation of the protection provided by common safety glasses from ultraviolet emissions in welding operation. *Am Ind. Hyg. Assoc. J.* 40:770-80 (1979).

Hughes EG: Cadmium fume. *Lancet* 2 (8195):641 (1980).

International Institute of Welding, Commission VIII Health and Safety: Formation of nitric-oxides in gas welding: Measures for the prevention of injuries to health. *Welding in the World* 18:164-71 (1980).

Jarzebski J: Thermal injuries of the middle ear in welders. *Med. Pr.* 30:457-60 (1979).

Jirasek L: Occupational exogenous siderosis of the skin. *Contact Dermatitis* 5:334-5 (1979).

Jones RN, Weill H: Occupational lung disease. *Respiratory Care* 23:989-98 (1978).

Juniczak T: Spirometric examinations of workers exposed to the harmful activity of dust and welding smokes at the H. Cegielski Metal Works in Poznan. *Pneumonol. Pol.* 47, 753-757 (1979).

Kalliomaki PL, Alanko K, Korhonen O, Mattsson T, Vaaranen V, Koponen M: Amount and distribution of welding fume lung contaminants among arc welders. *Scand. J. Work Environ. Health* 4:122-130 (1978).

Kalliomaki PL, Kalliomaki K, Kelha V, Korhonen O: Measure of lung contamination among mild steel and stainless steel welders. *Welding in the World* 18:67-72 (1980).

Kalliomaki K, Kalliomaki PL, Kelha V, Vaaranen V: Instrumentation for measuring the magnetic lung contamination of steel welders. *Ann. Occup. Hyg.* 23:175-84 (1980).

Kalliomaki PL, Sutinen S, Kelha V, Lakomaa E, Sortti V: Amount and distribution of fume contaminants in the lungs of an arc welder post mortem. *Br. J. Ind. Med.* 36:224-30 (1979).

Kaplan PD, Blackstone M, Richdale N: Cadmium oxide toxicity: macromolecular binding of cadmium, zinc, and copper in the fibrotic rat lung. Biol Implic Met Environ, Proc Annu Hanford Life Sci Symp, 15th 1977 (ERDA Symp Ser 42:77-97 (1977).

Kawada K, Iwano K, Nagai N, Kimura S: Experimental Studies on the Harmfulness of Basic Type Welding Electrodes and on their Improvement. International Institute of Welding, Commission VIII Document 197-64 (1964).

Keskinen H, Kalliomaki PL, Alanko K: Occupational asthma due to stainless steel welding fumes. *Clin. Allergy* 10:151-59 (1980).

Kersley: Arc flash and the contact lens wearer. *Brit. Med. J.* 251:639-40 (1977).

Kimura S: Coated Welding Electrode Containing Chromium. *US Patent* 4131784 7 pp (1978).

Kimura S, Kobayashi M, Godai T, Minato S: Investigation on chromium in stainless steel welding fumes. *Welding Journal* 58:195S-204S (1979).

King E: Cadmium fume. *Lancet* 2 (8195):641 (1980).

Knudsen I: The mammalian spot test and its use for the testing of potential carcinogenicity of welding fume particles and hexavalent chromium. *Acta Pharmacol. Toxicol.* (kbh) 47:66-70 (1980).

Kobayashi M, Sakai Y, Hayashi K: Development of fume controlled electrodes. *Welding in the World* 17:228-31 (1979).

Koshi K: Effects of fume particles from stainless steel welding on sister chromatid exchanges and chromosome aberrations in cultured Chinese hamster cells. *Ind. Health* 17:39-49 (1979).

Krieg AH: Welding fume problems—and how to solve them. *Construction Contracting* 61:26-30 (1979).

Kujawska A, Marek K: Radiological picture of pneumoconiosis in welders. *Med. Pr.* 30:79-84 (1979).

Kup W: Studies on occupational damage to upper respiratory tract mucosa. *Atomwegs-Lungenkr* 5:410-11 (1979).

Lam HF, Hewitt PJ, Hicks R: A study of pulmonary deposition, and the elimination of some constituent metals from welding fume in laboratory animals. *Ann. Occup. Hyg.* 21:363-73 (1979).

Langard S, Norserth T: Chromium in *Toxicology of Metals,* Vol II, EPA-600/1-77-022. Environmental Health Effects Research Series, U.S. Environmental Protection Agency, Research Triangle Park, N.C. 164 pp (1975).

Langard S. Norseth T: A cohort study of bronchial carcinomas in workers producing chromate pigments. *Brit. J. Ind. Med.* 32, 62-65 (1975).

Linde CJ: The effect of welding fumes on ocular readaptation time. *Scand. J. Work Environ. Health* 6:135-45 (1980).

Lippmann, M: Aerosols (Dusts, Mists, and Fumes). *CRC Press,* 121-71 (1972).

Loriot J, Felix M, Gacouin JC, Proteau J: Pulmonary siderosis of arc welders in the form of fatal diffuse interstitial fibrosis. *Arch. Mal. Prof.* 40:521-24 (1979).

Louria DB, Joselow MM, Browder AA: The human toxicity of certain trace elements. *Ann. Int. Med.* 76:307-319 (1972).

Lovsund P, Nilsson SEG, Lindh H, Obert P: Temperature changes in contact lenses in connection with radiation from welding arcs. *Scand. J. Work Environ. Health* 5:271-79 (1979 a).

Lovsund P, Oberg PA, Nilsson SEG: Influence on vision of extremely low frequency electromagnetic fields: Industrial measurements, magnetophosphene studies in volunteers and intraretinal studies in animals. *Acta Ophthalmol (Copenh)* 57:812-21 (1979 b).

Lanau FW: Ozone in Arc Welding. *Ann. Occup. Hyg.* 10, 175-188 (1967).

Macdonald EJ: Epidemiology of skin cancer, 1975. in Neoplasms of the Skin and Malignant Melanoma: A Collection of Papers Presented at the Twentieth Annual Clinical Conference on Cancer, 1975, at the University of Texas System Cancer Center, MD Anderson Hospital and Tumor Institute, Houston, Texas. Chicago, Year Book Medical Publishers Inc., 27-42 (1976).

Magnusson EJ: Welding fumes and productivity *Machinery and Production Engineering* 402-06 (1977).

Maxild J, Anderson M, Kiel P, and Stern RM: Mutagenicity of Fume Particles from Metal Arc Welding on Stainless Steel in the Salmonella/Microsome Test. *Mutat Res.* 56(3): 235-43 (1978).

McCarthy DS, Spencer R, Greene R, and Milic-Emili J: *Am. J. Med.* 52, 747 (1972).

McCord CP, Harrold GC, Meek SF: A Chemical and Physiological Investigation of Electric Arc Welding. III. Coated Welding Rods. *J. Ind. Hyg. Toxicol.* 23(5):200-15 (1941).

McDowell-Boyer LM: Estimated Radiation Doses from Thorium and Daughters Contained in Thoriated Welding Electrodes. *National Technical Information Service* 53 pp (1980 a).

McDowell-Boyer LM: Estimated Radiation Doses from Thorium and Daughters Contained in Thoriated Welding Electrodes. *Govt. Reports Announcements & Index* 80: p 2444 (1980 b).

McMillan GHG: Reflections on current European health and safety activities in relation to welding. *Welding Journal* 58 (12):21-25 (1979 a).

McMillan GHG: Studies of the health of welders in naval dockyards. *Ann. Occup. Hyg.* 21:377-92 (1979 b).

McMillan GHG: Welders' health examinations. *J. Soc. Occup. Med.* 29:87-92 (1979 c).

McMillan GHG: Health of welders, those in similar trades, their assistants and neighborhood workers. *Welding in the World* 18:115-18 (1980).

McMillan GHG, Heath J: The health of welders in naval dockyards: Acute changes in respiratory function during standardized welding. *Ann. Occup. Hyg.* 22:19-32 (1979).

Milham, Jr. S: Cancer mortality patterns associated with exposure to metals. *Ann. N.Y. Acad. Sci.* 271, 243-49 (1976).

Miller TM, Jones RC: An assessment of the fume characteristics of Australian electrodes for manual metal arc welding. *Australian Welding Research* 6:1-9 (1979).

Mlynarczyk W, Senczuk W: Correlation between changes in lungs and amino acid composition of blood plasma after exposure in industrial dusts. III. The usefulness of the results for diagnostic purposes. *Med. Pr.* 31:27-33 (1980).

Moreton J: The Generation and Control of Welding Fume. Bibliography: Particulate Fume and Pollutant Gases in the Welding Environment. *The Welding Institute* Publication 5510/19/77, Cambridge, Eng., p.74 (1977).

Moss CE: Development of welding/brazing radiation hazard criteria. *Toxicology Research Projects Directory* 5(3), (1980).

Moss CE, Gawenda MC: Optical Radiation Transmission Levels Through Transparent Welding Curtains, DHEW (NIOSH) Publication No. 78-176. Cincinnati, US Dept of Health Education and Welfare, Public Health Service, National Institute for Occupational Safety and Health, (1978).

Mutti A, Cavatorta A, Pedroni C, Borghi A, Giaroli C, Franchini I: The role of chromium accumulation in the relationship between airborne and urinary chromium in welders. *Int. Arch. Occup. Environ. Health* 43:123-133 (1979).

Naranjit D, Thomassen Y, Van Loon JC: Development of a procedure for studies of the chromium (III) and chromium (VI) contents of welding fumes. *Analytica Chemica Acta* 110:307-12 (1979).

National Institute of Occupational Safety and Health: Occupational Exposure to Chromium (VI). HEW Publication No. 76-129, 1975.

National Institute for Occupational Safety and Health: Safety and Health in Arc Welding and Gas Welding and Cutting, DHEW (NIOSH) Publication No. 78-138. Cincinnati, National Institute for Occupational Safety and Health, 50 pp (1978).

National Institute for Occupational Safety and Health: Documenting radiation effects. *Occup. Health Saf.* 48:54-57 (1979).

National Safety Council: Gas-shielded Arcs and Plasma-jet Torches, Data Sheet 552. Chicago, National Safety Council, 5 pp (1965).

National Safety Council: Soldering and Brazing, Data Sheet I-445-78. Chicago, National Safety Council, 7 pp (1978 a).

National Safety Council: Zinc and Zinc Oxide, Data Sheet 267. Chicago, National Safety Council, 2 pp (1978 b).

National Safety Council: Dusts, Fumes, and Mists in Industry, Data Sheet I-531-Rev80. Chicago, National Safety Council, 16 pp (1980).

Nauwald G: Knee joint changes among welders in ship-building *Beitr Orthop Traumatol* 27:181-7 (1980).

Oakley PJ: Ways to measure rates of fume emission. *Weld Design Fabrication* 52:74-75 (1979).

Occupational Safety and Health Administration: General Industry Safety and Health Standards, 29 CFR 1910. Washington, D.C., Government Printing Office, 820 pp (1976).

OSHA: Documentating radiation effects. *Occupational Health and Safety* 48(7):54-7 (1979).

Oxhoj H, Bake B, and Wilhelmsen L: *Scand. J. Resp. Dis.* 58:80 (1977).

Oxhoj H, Bake B, Wedel H, and Wilhelmsen L: Effects of electric arc welding on ventilatory lung function. *Arch. Environ. Health,* 34:211-217 (1979).

Pattee HE, Howden DG, Evans RM, and Martin DC: Improving the Welding Environment: Topical Report on Laboratory Investigation of Arc Welding Fumes and Gases, Submitted to the American Welding Society, Miami, FL, 1978.

Pinfold BE, Allum CJ, Darling DV, Nixon JH, Lees BH, King J: Underwater welding: a realistic assessment - some comments. *Welding and Metal Fabrication* 47:435-38 (1979).

Pollak VA, Romanchuk KG: The risk of retina damage from high intensity light sources. *Ann. Ind. Hyg. Assoc. J.* 41:322-27 (1980).

Post CT: How can industry combat welding fume problem? *Iron Age* 223:45-48 (1980).

Prat R, Nofre C, and Cier A: Effects of sodium hypochlorite, ozone and ionizing radiations on pyrimidine constituents of Escherichia Coli. Ann. Inst. Pasteur 114:595-607 (1968).

Press H: Concentrations of gases and fumes in plasma arc cutting *Welding in the World* 16:96-103 (1978).

Prosperi D: Method and apparatus for welding underwater. *Brit. Pat.* 1, 555, 340, 6 pp, Sept 12 (1979).

Rabenda A: Incidence of pneumoconiosis in welders in semiautomatic systems in the light of environmental studies. *Med Pr.* 31:239-53 (1980).

Riedinger U, Beaufils D: Problems poses par l'emission des fumees au cours du soudage a l'arc avec electrodes enrobees. *Soudage et Techniques Connexes* 331-36 (1978).

Romanchuk KG, Pollak V, Schneider RJ: Retinal burn from a welding arc, *Canad J Ophthal* 13:120-22 (1978).

Ross DS: Welder's Health - the respiratory system and welding. *Metal Construction,* 119-121 (1978).

Royle, H: Toxicity of chromic acid in the chromium plating industry (1). *Environ. Res.* 10, 39-53 (1975).

Sachsenmaler W., Siebs W. and Tan T.: Effects of ozone on mouse ascites tumor cells and on chick fibroblasis in tissue culture 67:113-126 (1965).

Samoilova LM, Kireev VI: Morphological Changes in the Lungs of Animals under the Influence of an Aerosol Formed during Welding and Surfacing of Cast Iron. *Gig. Tr. Prof. Zabol* (4): 53-54 (1975).

Schubert J: Manufacturer clears the air without heat loss. *Welding Journal* 58:54-56 (1979).

Sherr AE: Industrial safety glasses — an update. *J. Am. Optom. Assoc.* 51:129-38 (1980).

Shumay WC Jr.: What's happening with welding curtains? *Welding Design & Fabrication* 51:98-100 (1978).

Shutt RC: Controlled total copper in wire electrodes. *Welding Journal* 58:30-35 (1979).

Sloan D: Skin diseases - how industrial workers are vulnerable. Welding and Metal Fabrication 46:271-276 (1978).

Speight FY: Appendix: October 26, 1977 Letter from American Welding Society to OSHA. *Weld J* 57 (7): 22-23 (1978).

Stefanescu A, Nadudvary G, Borda M: Physical and chemical characterization of aerosols resulting from manual electric welding, with a view to establishing the occupational risks. *Constructia de Masini* 20:756-59 (1968).

Sterling WK: Optimal welding gas exhaust. *Zentralbl Arbeitsmed Arbeitsschutz Prophyl* 29:270 (1979).

Stern RM: A Chemical, Physical and Biological Assay of Welding Fumes. *The Danish Welding Institute Publication* 77.05 (1977).

Stern RM: Risks in chromium and nickel exposure for stainless steel workers. *Ann. Clin. Lab. Sci.* 8:501 (1978).

Suvorov IM, Gratsianskaia LN, Khaimovich ML, Tsirulnikova II, Zenkevich ES: Effect of a complex of occupational factors on the nature and characteristics of diseases in electric welders. *Vrach Delo* 7:98-102 (1980).

Task Group on Lung Dynamics, Committee II of the International Radiological Protection Commission: Deposition and retention models for internal dosimetry of the human respiratory tract. *Health Phys.* 12, 173-207 (1966).

Tengroth B, Vulcan J: Welding Light. Strahlenther 153(4):267-72 (1977).

Thomsen E, Stern RM: A simple analytical technique for the determination of hexavalent chromium in welding fumes and other complex matrices. *Scand. J. Work Environ. Health* 5:386-403 (1979).

Tierney MP: Analysis of nine injuries associated with maintenance and repair in metal and nonmental mines. *U.S. Dept. of the Interior Mining Enforcement and Safety Administration* 1977, pp 1-36.

Titus AC, Warren H, Drinker P: Electric Welding I. The Respiratory Hazard. *J. Ind. Hygiene* 17(4): 121-28 (1935).

Tola S, Karskela V: Occupational lead exposure in Finland v. shipyards and shipbreaking. *Scand. J. Work Environ. Health* 2:31-36 (1976).

Tossavainen A, Nurminen M, Mutanen P, Tolen S: Application of mathematical modelling for assessing the biological half-times of chromium and nickel in field studies. *Por. J. Ind. Med.* 37:285-291 (1980).

Ulfvarson U, Halline U, Bellander T, Scjoegran B, Swensson A: Welding Problems Connected with Work Environment. (Arbetsmiljoeproblem Vid Svetsning). 93 pp (1978).

Van Wagenen HD: Assessment of Selected Control Technology Techniques for Welding Fumes, DHEW (NIOSH) Publication No. 79-125. National Institute for Occupational Safety and Health 24 pp (1979).

Villaume JE, Wasti K, Liss-Suter D, and Hsiao S: Effects of Welding on Health. Report Submitted to *American Welding Society* (1978).

Virtamo M, and Tuomola S: Hexavalent Chromium Compounds in Welding Fumes. *International Institute of Welding,* Commission VIII Document 584-74. (1974).

Von Haam E, Groom JJ: The Pathology of Shielded Arc Welding. J. Ind. *Hygiene Toxicol* 23(2):55-57 (1941).

Weiler KJ: Nickel contact eczema caused by electric welding? *Derm Beruf Umweldt* 27:142-3 (1979).

Westgate SA: Safety and resistance welding. *Metal Construction* 11:371 (1979).

Weymueller CR: Keeping the welder healthy and safe. *Welding Design and Fabrication* 52:104-07 (1979).

Weymueller CR: On a clear day ... you can see across the shop: welding fume removal. *Welding Design and Fabrication* 52:122-26 (1979).

Weymueller CR: Submerged arc welding - making a comeback. *Welding Design and Fabrication* 52:49-54 (1979).

Weymueller CR: A fume-free shop - the ultimate goal. *Welding Design and Fabrication* 53:89-91 (1980).

Weymueller CR: How to win safety awards. *Welding Design and Fabrication* 53:131-33 (1980).

White LR, Jakobsen K, Ostgaard K: Comparative toxicity studies of chromium-rich welding fumes and chromium on an established human cell line. *Environ Res* 20:366-74 (1979).

Wilkinson ME: Welding at pressures greater than atmospheric pressures. *Bit Pat Application* 2,002, 276A, Feb 21, 1979, 6 pp.

Wilson JD, Stenzel MR, Lombardozzi KL, Nichols CL: Monitoring personnel exposure to stainless steel welding fumes in confined spaces at a petrochemical plant. *Am. Ind. Hyg. Assoc. J.* 42:431-36 (1981).

Wysocki K, Chrzanowska-Ansilewska K, Jurek D, Kersten E, Stanosz S: The behavior of various serum enzymes in welders with regard to the question of occupational liver damage *Z Aerztl Fortbild Jena* 74:818-19 (1980).

Zakhari S, Anderson RS: Effects of Welding on Health II. An update report submitted to the *American Welding Society,* 1980.

Zelac RE, Cromroy HL, Bolch, Jr WE, Donavant BG, Bevis HE: Chromosomal aberrations induced in Chinese hamster lymphocytes. *Environ. Res.* 4, 262 (1971).